METHODS DEVELOPMENT

ALLE

B

MW00843339

Immunohistochemistry: Basics and Methods

Igor B. Buchwalow · Werner Böcker

Immunohistochemistry: Basics and Methods

Springer

Prof. Dr. Igor B. Buchwalow
Prof. Dr. Werner Böcker
Gerhard-Domagk-Institut für Pathologie
Domagkstr. 17
48149 Münster
Germany

New addresses:

Prof. Dr. Igor B. Buchwalow
Institute for Hematopathology
Fangdieckstrasse 75a
22547 Hamburg
Germany
buchwalow@pathologie-hh.de;
buchwalow@hotmail.de

Prof. Dr. Werner Böcker
Consultation and Reference Center
for Gyneco- and Breast-Pathology
Fangdieckstrasse 75a
22547 Hamburg
Germany
boecker@pathologie-hh.de;
werner.boecker@gmail.com

ISBN: 978-3-642-04608-7 e-ISBN: 978-3-642-04609-4
DOI 10.1007/978-3-642-04609-4
Springer Heidelberg Dordrecht London New York

Library of Congress Control Number: 2009938017

Cover Illustration: Double immunolabeling of cytokeratins in a duct wall of the human mammary gland, see Chap. 8, Fig. 3 (photo by Igor B. Buchwalow, Gerhard-Domagk-Institut für Pathologie, Münster, Germany)

Cover design: WMXDesign GmbH, Heidelberg, Germany

Printed on acid-free paper

Springer is part of Springer Science+Business Media (www.springer.com)

Preface

Combining two different scientific disciplines – morphology and immunochemistry – immunohistochemistry has developed as an important instrument in research and clinical pathology. A basic understanding of underlying principles and potential problems is unavoidable if you want to be successful in your use of immunohistochemistry, as well as in getting your papers published and your research grants funded.

While many excellent texts and monographs exist which cover various aspects of immunohistochemistry, the lack of a concise comprehensive guide to using these methods was a major motivation for writing this book. Our intention was to create an easy-to-read and focused resource based on state-of-the-art information for a broad audience ranging from students and technical assistants to experienced researchers. This handbook has a concise format, with protocols and instructions for methods immediately following the short introductory theoretical material in each chapter. Being conscious of the growing role of Internet as an information source, we have found it reasonable in many cases to substitute citing books and journal publications with corresponding Internet websites. Where possible, commercial sources of reagents, kits, and equipment are listed throughout the text instead of in a separate index. Though each chapter is small and introductory, this handbook itself is self-sufficient and provides a comprehensive look at the principles of immunohistochemistry. For readers wanting further depth of knowledge, each chapter is backed up by a short list of carefully selected original articles.

During the last decade, pioneering efforts of histochemists have led to an immense improvement in the reagents and protocols. The researcher is urged always to determine the reason for every method and step before doing it. This handbook is intended to help readers to avoid troubles in the choice of an adequate method, which happens when using standard textbooks. For this handbook, we carefully selected established methods and easy-to-adopt protocols, paying attention to modern developments in immunohistochemistry, such as antigen retrieval, signal amplification, the use of epitope tags in immunohistochemistry, multiple immunolabeling or diagnostic immunohistochemistry. Each of the methods described in this handbook was proved by the authors; many of these methods are routinely used in daily practice in their institute. All the practical methods advocated are clearly described, with accompanying tables, and the results obtainable are illustrated with colour micrographs.

Acknowledgements We thank Vera Samoilova for the perfect technical assistance and other colleagues from the Münster University Clinic for sharing probes and reagents.

Igor B. Buchwalow and Werner Böcker
Münster

Contents

Chapter 1
Antibodies for Immunohistochemistry

The first use of the term *Antikörper* (the German word for *antibody*) occurred in a text by Paul Ehrlich (Fig. 1.1) in the conclusion of his article "Experimental Studies on Immunity," published in October 1891. Paul Ehrlich was born in 1854 in Strehlen (the German Province of Silesia, now in Poland). As a schoolboy and student of medicine he was interested in staining microscopic tissue substances. In his dissertation at the University of Leipzig, he picked up the topic again ("Contributions to the Theory and Practice of Histological Staining," Beiträge zur Theorie und Praxis der histologischen Färbung) (http://en.wikipedia.org/wiki/Paul_Ehrlich). In 1903, Paul Ehrlich published the ever first comprehensive text-book describing histological and histochemical staining techniques ("Encyclopedia of Microscopical Technique," Enzyklopädie der Mikroskopischen Technik). His first immunological studies were begun in 1890 when he was an assistant at the Institute for Infectious Diseases under Robert Koch. In 1897, Paul Ehrlich proposed his theory for antibody and antigen interaction, when he hypothesized that receptors on the surface of cells could bind specifically to toxins — in a "lock-and-key interaction" — and that this binding reaction was the trigger for the production of antibodies. He shared the 1908 Nobel Prize with Mechnikoff for their studies on immunity (http://en.wikipedia.org/wiki/Antibody).

Albert H. Coons (Fig. 1.2) was the first who attached a fluorescent dye (fluorescein isocyanate) to an antibody and used this antibody to localize its respective antigen in a tissue section. The concept of putting a visible label on an antibody molecule appeared both bold and original. His initial results were described in two brief papers in the early 1940s (Coons et al. 1941, 1942), but the research was halted while he joined the army and spent the next 4 years in the South Pacific. His later studies (Coons and Kaplan 1950) contributed immensely to the use of the fluorescent antibody method in a wide variety of experimental settings. In our time, the use of antibodies to detect and localize individual or multiple antigens in situ has developed into a powerful research tool in almost every field of biomedical research (http://books.nap.edu/html/biomems/acoons.pdf).

I.B. Buchwalow and W. Böcker, *Immunohistochemistry: Basics and Methods*,
DOI 10.1007/978-3-642-04609-4_1, © Springer-Verlag Berlin Heidelberg 2010

Fig. 1.1 Paul Ehrlich

Fig. 1.2 Albert H. Coons
(Courtesy of the Harvard
Medical School Countway
Library)

1.1   Structure of Antibodies

An antibody or immunoglobulin (Ig) is a glycoprotein used by the immune system
to identify and neutralize substances foreign to the body, such as bacteria and
viruses and other infectious agents, known as antigens. Some immunologists will
argue that the word immunoglobulin covers more than just antibodies but we will
not complicate matters by going in to the details here. In mammals, there are five
classes of antibodies: IgG, IgA, IgM, IgE and IgD. For immunoassays, two Ig
classes are of importance — IgG and IgM.

IgM eliminates pathogens in the early stages of B cell mediated immunity before
there is sufficient IgG. The IgG molecule has two separate functions; to bind to the
pathogen that elicited the response, and to recruit other cells and molecules to
destroy the antigen. Generated during the secondary immune response and produced by B lymphocytes, IgG molecules provide the majority of antibody-based
immunity against invading pathogens. IgG is one of the most abundant proteins in
serum of adult animals and constitutes approx. 75% of serum immunoglobulins.
Therefore, it is primarily used in the experiments for production of antibodies. IgG
antibodies are further divided into four subclasses (also referred to as isotypes,
IgG1, IgG2, IgG3 and IgG4).

The classical "Y" shape of the IgG molecule (MW ~150 kD) is composed of four
polypeptide chains — two identical light chains (each has a molecular weight of
~25 kD), and two identical heavy chains (each has a molecular weight of ~50 kD)

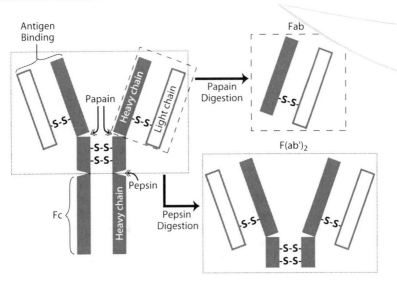

Fig. 1.3 Schematic representation of an antibody molecule. Adapted from http://probes. invitrogen.com/handbook/boxes/0439.html

— which are connected by disulfide bonds (see Fig. 1.3). Each end of the forked portion of the "Y" on the antibody is called the *Fab* (*Fragment, antigen binding*) region. Antigen specific Fab "arms" are responsible for antigen binding. A Fab fragment is comprised of one light chain and the segment of heavy chain on the N-terminal side. The light chain and heavy chain segments are linked by interchain disulfide bonds. The *Fc* "tail" (*Fragment, crystallizable*) is a region of an antibody composed of two heavy chains on the C-terminal side. Fc has many effector functions (e.g., binding complement, binding to cell receptors on macrophages and monocytes, etc.) and serves to distinguish one class of antibody from another (Harlow and Lane 1999).

Fc and Fab fragments can be generated through digestion by enzymes. An antibody digested by papain yields three fragments, two Fab fragments and one Fc fragment. Pepsin attacks the antibody molecule on the heavy chain below disulfide bonds connection, cleaving the antibody into a $F(ab')_2$ fragment and a Fc fragment. Along with the whole antibody molecules, Fab, $F(ab')_2$ and Fc fragments are useful tools in immunoassays. Bivalent $F(ab')_2$ fragments can be used as secondary antibodies to avoid the nonspecific binding to Fc receptors in tissue sections, especially on the cell surface. Unconjugated monovalent Fab fragments can be used to convert the primary antibody into a different species. Labeled monovalent Fab fragments are used for haptenylation of primary antibodies in double or multiple immunostaining when primary antibodies belong to the same species (see Sect. 8.2). The Fc fragment serves as a useful "handle" for manipulating the antibody during most immunochemical procedures. Antibodies are usually labeled in the Fc region for immunohistochemical labeling.

Since IgG is the most abundant immunoglobulin in serum (http://probes. invitrogen.com/handbook/boxes/0439.html), it is primarily used in the production of antibodies for immunoassays. IgM accounts for approximately 10% of the immunoglobulin pool, and is also used in the production of antibodies, but to a lesser extent than IgG. The IgM molecule contains five or occasionally six "Y"-shaped subunits covalently linked together with disulfide bonds. The individual heavy chains have a molecular weight of approximately 65,000 and the whole molecule has a molecular weight of 970,000 (Harlow and Lane 1999).

1.2 Polyclonal Antibodies

Polyclonal antibodies are produced by immunization of suitable animals, usually mammals. An antigen is injected into the animal, IgG specific for this antigen are produced by B lymphocytes as immune response, and these immunoglobulins are purified from the animal's serum. Thereby, specific IgG concentrations of approximately 1–10 mg/ml serum can be obtained (see Table 4.1). Antibodies produced by this method are derived from different types of immune cells, and hence are called polyclonal. Animals usually used for polyclonal antibody production include goats, horses, guinea pigs, hamsters, mice, rats, sheep and chickens. Rabbit and mice are the most commonly used laboratory animals for this purpose. Mice are, however, mostly used for production of monoclonal antibodies (see Sect. 1.3). Larger mammals, such as goats or horses, are used when large quantities of antibodies are required. Many investigators favor chickens because chickens transfer high quantities of IgY (IgG) into the egg yolk, and harvesting antibodies from eggs eliminates the need for the invasive bleeding procedure. One week's eggs can contain ten times more antibodies than the volume of rabbit blood obtained from one weekly bleeding (http://en.wikipedia.org/wiki/Polyclonal_antibodies).

1.3 Mouse Monoclonal Antibodies

Whereas polyclonal antibodies are multiple antibodies produced by different types of immune cells that recognize the same antigen, monoclonal antibodies are derived from a single cell line (also referred to as clone). In monoclonal antibody technology, tumor cells that can replicate endlessly are fused with mammalian cells that produce an antibody. The result of this cell fusion is called a "hybridoma," which will continually produce antibodies. Monoclonal antibodies are identical because they were produced by immune cells that all are clones of a single parent cell. Production of monoclonal antibodies requires immunizing an animal, usually a mouse, obtaining immune cells from its spleen, and fusing the cells with a cancer cell (such as cells from a myeloma) to make them immortal, which means that they will grow and divide indefinitely (see Fig. 1.4). In fermentation chambers, antibodies can be produced on a larger scale. Nowadays, bioengineering permits production of antibodies also in plants.

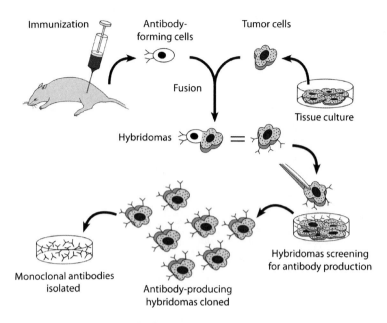

Fig. 1.4 Monoclonal antibody production. Adapted from http://www.accessexcellence.org/RC/VL/GG/monoclonal.html

Monoclonal antibodies can be produced not only in a cell culture but also in live animals. When injected into mice (in the peritoneal cavity, the gut), the hybridoma cells produce tumors containing an antibody-rich fluid called ascites fluid. Production in cell culture is usually preferred, as the ascites technique may be very painful to the animal and if replacement techniques exist, may be considered unethical. The process of producing monoclonal antibodies described above was invented by Georges Köhler, César Milstein, and Niels Kaj Jerne in 1975; they shared the Nobel Prize in Physiology or Medicine in 1984 for the discovery (http://en.wikipedia.org/wiki/Antibody).

1.4 Rabbit Monoclonal Antibodies

Monoclonal antibodies have traditionally been produced in a mouse and to a lesser extent in a rat. Cloning techniques produce highly specific antibodies. However, polyclonal antibodies produced by rabbits are known to have a higher affinity than antibodies raised in mice, which means — to put it simply — a better antigen recognition. Additionally, the rabbit immuno-response recognizes antigens (epitopes) that are not immunogenic in mice. It is often for this reason that rabbit polyclonal antibodies are developed and used in many research and even some diagnostic applications. In order to combine benefits of better antigen recognition

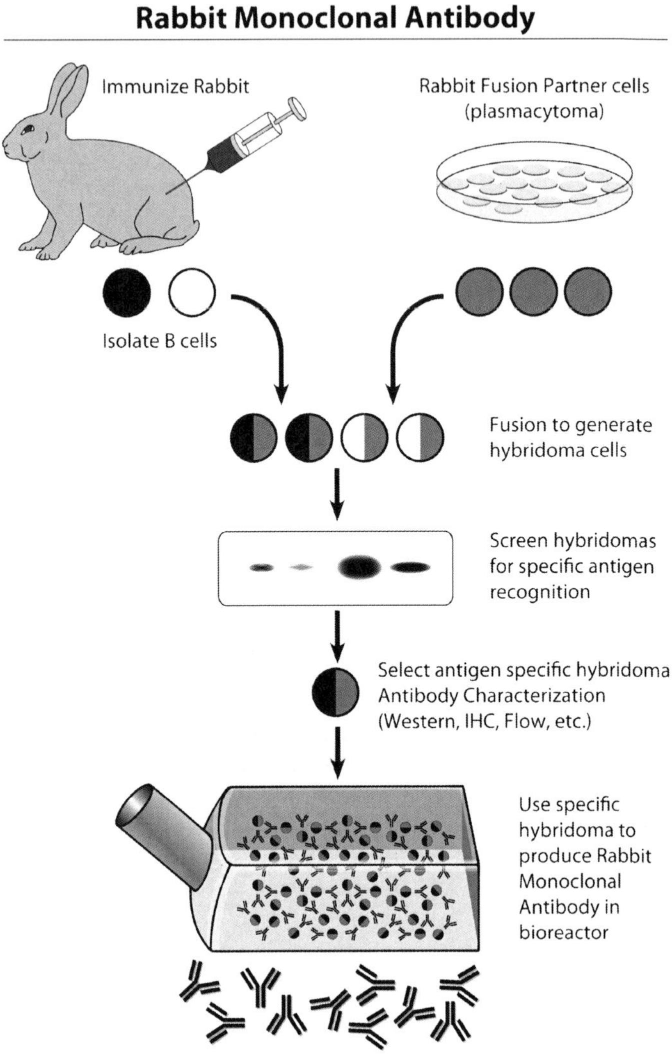

Fig. 1.5 The general outline that Epitomics uses for producing a rabbit monoclonal antibody

by polyclonal antibodies with the specificity and consistency of a monoclonal antibody, Epitomics (http://www.epitomics.com/) developed a proprietary unique method for making monoclonal antibodies from rabbits (Fig. 1.5). Rabbit monoclonal antibodies are now available also from Lab Vision Corporation (http://www.labvision.com/rabmab/Rabmab.cfm) and Bethyl Laboratories (http://www.bethyl.com/default.aspx).

The basic principle for making the rabbit monoclonal antibody is the same as for mouse monoclonals. Rabbit fusion partner cells can fuse to rabbit B-cells to create the rabbit hybridoma cells. Hybridomas are then screened to select for clones with specific and sensitive antigen recognition, and the antibodies are characterized using a variety of methods. The resulting rabbit monoclonal antibody has ten times the affinity of the mouse antibody, thus resulting in a more specific and much more sensitive antibody.

1.5 Protein A and Protein G in Immunohistochemistry

Protein A and protein G are bacterial proteins that bind with high affinity to the Fc portion of various classes and subclasses of immunoglobulins from a variety of species. Protein A is a constituent of the *Staphylococcus aureus* cell wall. Protein A was introduced in immunocytochemistry for the localization of different antigens at the light and electron microscope levels in the 1970s (Roth et al. 1978). Later, another protein isolated from the cell wall of group G streptococcal strains, known as protein G, was found to display similar IgG-binding properties. Like protein A, protein G has a high affinity for IgGs from various mammalian species (Table 1.1). Protein G was, however, found to interact with a broader range of polyclonal and monoclonal antibodies from various mammalian species with higher avidity (Bendayan and Garzon 1988). Protein A/G is a genetically engineered 50,460 Da protein that combines IgG binding domains of both protein A and protein G. Protein A/G contains four Fc binding domains from protein A and two from protein G. It is a gene fusion product secreted from a non-pathogenic form of *Bacillus*. Protein A/G binds to all subclasses of mouse IgG but does not bind mouse IgM. Mouse monoclonal antibodies commonly have a stronger affinity to the chimeric protein A/G than to either protein A or protein G.

The use of protein A and protein G in immunohistochemistry is based on the same principle as that using secondary anti-IgG antibodies in the indirect two-step approach. In the first step, an antigen–IgG complex is formed, which is then revealed in the second step by incubation with labeled protein A or protein G.

Table 1.1 Binding profiles of protein A, protein G and protein A/G. Adapted from http://www.2spi.com/catalog/chem/gold_conjugate.html)

Primary IgG	Protein A	Protein G	Protein A/G
Chicken	+/−	+	+/−
Goat	+/−	++	+++
Guinea pig	++	++	++
Mouse IgG_1[a]	+/−	++	+/−
Mouse (other subclasses)	++	++	++
Rabbit	+++	+++	+++
Rat	+/−	+	+++
Sheep	+/−	++	+++

[a]IgG_1 binds best to protein A at a pH of 8–9. + Moderate binding, +++ Strong binding, − Weak or no binding

Molecular Probes offers several Alexa Fluor dye conjugates of protein A and G, providing bright and highly photostable conjugates with green to deep-red fluorescence (For more details see http://prob es.invitrogen.com/). Due to their significantly lower molecular weight (MW~40 kD) than the IgG molecule, protein A and G are also successfully exploited in electron microscopical immunohistochemistry. Once tagged with colloidal gold, the protein A- and protein G-gold complexes are able to react with various monoclonal and polyclonal primary antibodies, making possible the detection of a variety of tissue antigens directly under electron microscope (see Chap. 12).

References

Bendayan M, Garzon S (1988) Protein G-gold complex: comparative evaluation with protein A-gold for high-resolution immunocytochemistry. J Histochem Cytochem 36:597–607

Coons AH, Creech H, Jones R (1941) Immunological properties of an antibody containing a fluorescent group. Proc Soc Exp Biol Med 47:200–202

Coons AH, Creech H, Jones R, Berliner E (1942) The demonstration of pneumococcal antigen in tissues by the use of fluorescent antibody. J Immunol 45:159–170

Coons AH, Kaplan MH (1950) Localization of antigen in tissue cells II: Improvements in a method for the detection of antigen by means of fluorescent antibody. J Exp Med 91:1

Harlow E, Lane D (1999) Using Antibodies: a Laboratory Manual. Cold Spring Harbor Laboratory Press, NY

Roth J, Bendayan M, Orci L (1978) Ultrastructural localization of intracellular antigens by the use of protein A-gold complex. J Histochem Cytochem 26:1074–1081

Chapter 2
Antibody Labeling and the Choice of the Label

In order for the antigen–antibody immunoreaction to be seen in the microscope, the antibody must take a label – enzyme, fluorophore, colloidal gold or biotin. An enzyme label can be visualized in the light microscope by means of enzyme histochemical methods via chromogenic reactions (see Sect. 2.3). A fluorophore label can be directly visualized in a fluorescent microscope (see Sect. 2.4). Electron-dense labels such as colloidal gold are visible in the electron microscope without further treatment (see Sect. 12.1). Biotin label can be exploited in light, fluorescence and electron microscopy in combination with ABC technique (see Sect. 6.2.1). Like biotin, some other haptens, such as digoxigenin (DIG) or dinitrophenol (DNP), can also be coupled to antibodies. For their visualization, enzyme- or fluorophore-conjugated secondary antibodies are affordable.

2.1 Covalent Labeling of Antibodies

The methods for covalent labeling antibodies are beyond the scope of this book, and therefore will be discussed here only briefly. Covalent labeling of antibodies is not a routine procedure in the majority of laboratories, and requires relatively large quantities of purified antibody. One should also be familiar with how to use a desalting column, how to purify conjugates by dialysis or by gel filtration and how to take absorbance spectra (http://www.antibodybeyond.com/techniques/ab-conjugation.htm). For details, see the work of Harlow and Lane (1999). Labels can be chemically introduced into antibodies via a variety of functional groups on the antibody using appropriate group-specific reagents (http://www.drmr.com/abcon/). Many fluorophores bear a reactive group through which they readily combine covalently with antibodies in alkaline solutions (pH 9–10) (http://en.wikipedia.org/wiki/Cyanine). Labeling with an enzyme requires an additional large molecule such as glutaraldehyde to cross-link the enzyme to the antibody (Sternberger 1986).

The most widely applied principle is haptenylation of amino groups via N-hydroxysuccinimide esters (NHS-ES). For convenient protein-labeling procedures,

I.B. Buchwalow and W. Böcker, *Immunohistochemistry: Basics and Methods*,
DOI 10.1007/978-3-642-04609-4_2, © Springer-Verlag Berlin Heidelberg 2010

a number of haptens (various fluorophores, enzymes, biotin, DIG, etc.) are now commercially available as activated NHS-ES (Pitt et al. 1998) from various vendors (Sigma, Molecular Probes, Pierce and many others). By this procedure, the hapten moiety is introduced into the protein molecule via the reaction of the activated ester with primary amino groups of the antibody. However, many commercially available "purified" antibodies are formulated with BSA as a stabilizer, and amino groups present on the BSA molecule may reduce the efficiency of conjugation. Attempts to remove BSA from small quantities of antibody (e.g., 100 µg pack size) inevitably lead to substantial losses of antibody. To avoid this problem, Innova Biosciences (http://www.innovabiosciences.com/) developed a novel patented Lightning-Link^{TM} conjugation system, by which good quality conjugates can be prepared in the presence of surprisingly high concentrations of BSA. This technology requires only microgram quantities of antibody and just 30 s hands-on time to prepare antibody conjugates, even in the presence of BSA, without the otherwise obligatory preliminary step of antibody purification. To the time of this writing, we have not yet had the opportunity to evaluate this apparently promising conjugation system.

Should the antibody purification nevertheless be desirable, one can use the Ab-Select^{TM} Purification Kit (also from Innova Biosciences, http://www.innovabiosciences.com/products/abselect.php) that quickly removes the contaminants, such as BSA, glycine, tris or azide. The AbSelect method involves capture of the antibody on protein A resin and the removal of unwanted substances by a simple wash procedure. The purified product is then eluted and neutralized.

2.2 Non-Covalent Labeling of Primary Antibodies with Labeled Fab Fragments

A flexible and economic alternative to covalent labeling of antibodies was recently reported by Brown et al. (2004). Originally developed for using mouse monoclonal antibodies when working with mouse tissues, this method can be successfully applied for labeling small aliquots of antibodies (10 µg or even less). In this method, primary antibodies are non-covalently labeled with a reporter molecule in vitro using as a bridge monovalent Fab fragments that recognize both the Fc and F(ab')$_2$ regions of IgG. Simple mixing of labeled monovalent Fab fragments with an unconjugated primary antibody (see Fig. 2.1, Step 1) rapidly and quantitatively forms a labeling complex (labeled Fab fragment attached to primary antibody). After absorption of unbound monovalent Fab fragments with excess serum from the same species as the primary antibody (see Fig. 2.1, Step 2), the resultant complexes can be used for immunostaining both in multiple immunolabeling (see Sect. 8.2), and when applied to tissue samples homologous to the primary antibody species (see Sect. 9) avoiding the cross-reaction of the secondary antibody with endogenous immunoglobulins.

Complexation of primary antibodies with labeled Fab fragments

Step 1: Mixing of unconjugated primary antibodies with labeled monovalent Fab fragments

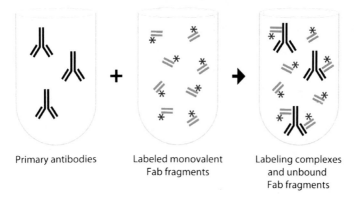

| Primary antibodies | Labeled monovalent Fab fragments | Labeling complexes and unbound Fab fragments |

Step 2: Absorption of unbound labeled monovalent Fab fragments with excess serum from the same species as the primary antibody

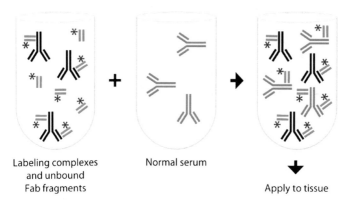

Labeling complexes
and unbound
Fab fragments

Normal serum

Apply to tissue

Fig. 2.1 Schematic representation of the procedure for generating and using primary antibody–Fab fragment complexes

For single-color immunolabeling, it is usually not necessary to remove free uncomplexed Fab fragments. However, in applications that involve multiple antibodies or when using primary antibody raised in the same species as the specimen to be immunostained, adsorption of the residual free Fab fragments is essential to avoid cross-reactivity. This approach is especially useful in multiple immunolabeling using primary antibodies belonging to the same species or even to the same IgG isotype (Buchwalow et al. 2005). Such in-house-made Fab fragment complexes of different primary antibodies bearing different labels can be used for simultaneously labeling multiple antigens in the same sample in a single incubation, avoiding the erroneous cross-reactions between different primary and secondary antibody pairs.

Protocol for generating primary antibody–Fab fragment complexes (Adapted from Brown et al. 2004)

1. Incubate unconjugated primary antibodies with labeled* monovalent Fab fragments directed against the given IgG species at a ratio** of 1:2 (weight for weight, based on concentration data supplied by manufacturers) in a small volume (e.g., in 10 µl or more, typically 1 µg of primary antibody in 10 µl) of staining buffer in a microcentrifuge tube for 20–30 min at room temperature.
2. Dilute the resultant primary antibody–Fab fragment complexes in staining buffer containing 10–20% normal serum from the same species as the primary antibody to give the concentration of about 1–10 µg/ml of the primary antibody, and incubate for 15–30 min at room temperature to block unbound labeled monovalent Fab fragments***.
3. Further dilute (if required) the resultant primary antibody–Fab fragment complexes to optimal working concentration (usually about 1–5 µg/ml) in staining buffer containing 10% normal serum and then apply to the sample for 30–60 min at room temperature and proceed further with your standard immunostaining protocol.
4. Prior to application of the resultant primary antibody–Fab fragment complexes to tissue sections, preincubate them with 10% normal mouse serum for 5–10 min.
   Notes: *The label may be biotin or any fluorophore. For antibody biotinylation, we use biotin-SP-conjugated AffiniPure Fab Fragment Goat Anti-Mouse IgG (H+L) (Jackson ImmunoResearch Labs, Code Number: 115-067-003).
   **Primary antibody to Fab fragment ratios of 1:2 (weight for weight, based on concentration data supplied by manufacturers) typically produces optimal results with mouse or rat primary monoclonal antibodies, as well as with rabbit or goat primary polyclonal antibodies.
   ***This resultant mixture of primary antibody–Fab fragment complexes containing blocked labeled monovalent Fab fragments can be stored for a few days at +4°C until use.

Jackson ImmunoResearch Laboratories Inc (http://www.jacksonimmuno.com/) sells all the components to develop in-house such staining. Suitable but rather expensive kits exploiting monovalent Fab fragments bearing fluorophore or biotin label have also been developed commercially, such as Animal Research Kit (ARK) by Dako (http://www.dakousa.com/) and Zenon Labeling Kits by Molecular Probes (http://probes.invitrogen.com/products/zenon/).

For double immunolabeling with primary antibodies from the same host species, it is not necessary to resort to labeling with monovalent Fab fragments when primary antibodies from the same host species are different isotypes (subclasses) of IgG, for instance such as mouse IgG1 and mouse IgG3. In these cases, isotype-specific antibodies may be used to distinguish between the two primary antibodies (see Sect. 8.2). For alternative methods see also website http://www.jireurope.com/ technical/fab-blok.asp.

2.3 Enzyme Labels for Light Microscopy

Enzyme labels can be visualized in the light microscope using the brightfield illumination modus. Brightfield illumination is the most widely used observation mode in optical microscopy for the past 300 years. For an overview of the elements that comprise light microscopes and theories behind important concepts in light microscopy, such as magnification, image formation, objective specifications, condensers, Köhler illumination, etc., the reader is referred to the websites: http://micro.magnet.fsu.edu/primer/anatomy/anatomy.html and http://en.wikipedia.org/wiki/Microscopy (see Chap. 14).

Enzyme labels are usually coupled to secondary antibodies or to (strept)avidin. The latter is used for detection of biotinylated primary or secondary antibodies in ABC methods (see Sect. 6.2.1). Enzyme labels routinely used in immunohistochemistry are horseradish peroxidase (HRP) and calf intestinal alkaline phosphatase (AP). Glucose oxidase from *Aspergillus niger* and *E. coli* β-galactosidase are only rarely applied.

Virtually all enzymes are proteins, and the molecules upon which they act are known as their substrates. The enzyme label can be visualized by means of enzyme histochemical methods via chromogenic reactions. "Chromogenic" means "producing color." In enzyme-histochemical chromogenic reactions, a soluble colorless substrate is converted into a water-insoluble colored compound either directly or in a coupled reaction. Histochemical detection of peroxidase is based on the conversion of aromatic phenols or amines, such as diaminobenzidine (DAB), into water-insoluble pigments in the presence of hydrogen peroxide (H_2O_2). Alkaline phosphatase catalyzes the hydrolysis of a variety of phosphate-containing substances in the alkaline pH range. The enzymatic activity of alkaline phosphatase can be localized by coupling a soluble product generated during the hydrolytic reaction with a "capture reagent," producing a colored insoluble precipitate.

For visualization of peroxidase label in tissue sections, Vector Laboratories offers the traditional substrates DAB and 3-amino-9-ethyl carbazole (AEC) as well as several proprietary substrates, producing colors as listed in Table 2.1.

AEC is soluble in alcohol and clearing agents, and must be mounted in aqueous mounting media. All other substrates are not soluble in alcohol or clearing agents. They may be dehydrated, cleared, and permanently mounted. These substrates can be used as single labels or to introduce multiple colors in a tissue section as shown for instance in Fig. 6.1 (see Chap. 7).

The alkaline phosphatase substrates form precipitates based on either reduction of tetrazolium salts or on the production of colored diazo compounds. Substrates Vector Red, Vector Black, Vector Blue, and BCIP/NBT available from Vector Laboratories produce reaction products which are red, black, blue and purple/blue

Table 2.1 Enzyme substrates from Vector Laboratories (http://www.vectorlabs.com/)

Enzyme substrates	Cat #	Color
ImmPACT™ DAB	SK-4105	Brown
Diaminobenzidine (DAB)	SK-4100	Brown
DAB + Ni2+	SK-4100	Gray/Black
Vector® VIP	SK-4600	Purple
Vector® SG	SK-4700	Blue-Gray
Vector® NovaRED™	SK-4800	Dark Red
TMB	SK-4400	Blue
3-amino-9-ethyl carbazole (AEC)	SK-4200	Red

respectively. The Vector Red, Vector Black, and BCIP/NBT reaction products can be permanently mounted in non-aqueous media. Vector Blue can also be permanently mounted in non-xylene mounting media if xylene substitutes are used to clear tissue sections. Development times may differ depending upon the level of antigen, the staining intensity desired or the substrate kit used. Generally, the Vector alkaline phosphatase substrate kits should be developed for 20–30 min. BCIP/NBT (unlike the other alkaline phosphatase substrates) will provide increased levels of sensitivity if the substrate incubation time is lengthened (up to 24 h). Importantly, the Vector Red reaction product is a highly fluorescent, bright red precipitate when viewed with rhodamine or Cy3 filter systems. Vector Red fluorescence may also be visible with FITC or DAPI filter systems using broad band emission filters. Up to three colors can be effectively introduced into a section to localize three antigens in different cells even using the same species of primary antibody or the same biotinylated secondary antibody.

Convenient ready-to-use enzyme substrate kits for visualization of enzyme labels are also available from DAKO (http://www.dako.com/), DCS Innovative Diagnostik Systeme (http://www.dcs-diagnostics.com/) and from other vendors. Commercial ready-to-use enzyme substrates may be excellent for what you want to do; however, they are not very cheap and in most cases their formulation is a trade secret. Apart from monetary savings, there are other good reasons to mix your own: you know what you have got. If you have time and inclination to prepare chromogen substrate solutions yourself, good and thorough procedures may be found in handbooks of Polak and Van Noorden (1997) and Lojda et al. (1976) or on the websites: http://www.ihcworld.com/chromogen_substrates.htm; http://www.ihcworld.com/_books/Dako_Handbook.pdf). For a complete review of enzyme histochemical methods, the reader is referred to handbooks of Lojda et al. (1976) and Van Noorden and Frederiks (1993). Useful information about enzyme chromogenic systems can also be derived from manufacturers' catalogs.

Basic protocol for immunoenzymatic staining
  1. Deparaffinize and rehydrate tissue sections. Rinse in distilled water for 5 min.
  2. Antigen retrieval: place sections in a Coplin jar with antigen retrieval solution of choice (e.g., 10 mM citrate acid, pH 6) and heat at 90°–110°C (depending on

tissue) in a microwave, steamer, domestic pressure cooker or autoclave (see Sect. 6.1.1).

3. Washes and dilutions: wash sections in PBS or TBS for 2 × 3 min. PBS can be used for all washes and dilutions. When working with AP-labeled secondary antibodies, TBS must be used.

4. Circle sections with a hydrophobic barrier pen (e.g., Dako Pen, #S2002) and proceed with immunohistochemical staining protocol. Do not allow sections to dry for the remaining procedure.

5. Blocking step 1: to block endogenous enzyme activity or endogenous biotin, use the corresponding blocking system (see Chap. 5) and wash in PBS or TBS for 2 × 3 min.

6. Blocking Step 2: to block endogenous Fc receptor, incubate sections for 20 min with PBS or TBS containing 5% normal serum of species in which the secondary antibodies were raised.

7. Primary antibodies: blot excess blocking solution from sections, and incubate for 60 min at room temperature or over night at +4°C with a correspondingly diluted primary antibody. Wash sections in PBS or TBS for 2 × 3 min.

8. Secondary antibodies: incubate sections for 30–60 min at room temperature with a HRP- or AP-labeled secondary antibody* raised against the corresponding IgG of the primary antibody. Wash sections in a buffer recommended for the corresponding enzyme substrate development for 2 × 3 min.

9. Substrate: incubate sections with an appropriate enzyme substrate** until optimal color develops. Wash sections first in the same buffer and thereafter in distilled water.

10. Counterstaining and mounting: counterstain nuclei (if desired) and coverslip using the appropriate protocol for aqueous or permanent mounting, depending on the chemical nature of the developed chromogen (consult the insert leaflet of the manufacturer).

11. All incubations are at room temperature unless otherwise noted.

   Notes: *When using biotin-labeled secondary antibodies instead of enzyme-labeled antibodies, you have first to detect biotin with enzyme-labeled (strept) avidin and proceed further with the Substrate Step (9). Do not add normal serum, non-fat dried milk, culture media or other potential sources of biotin to (strept)avidin-containing reagents. This may result in reduced sensitivity.

   **Solutions containing sodium azide or other inhibitors of peroxidase activity should not be used in diluting the peroxidase substrate.

2.4 Fluorophore Labels for Fluorescence Microscopy

In contrast to brightfield microscopy, which uses specimen features such as light absorption, fluorescence microscopy is based on the phenomenon in which absorption of light by fluorescent molecules called fluorescent dyes or fluorophores (known also as fluorochromes) is followed by the emission of light at longer wavelengths, usually in the visible region of the spectrum. In other words, fluorophores are getting themselves the light source. They can be visualized in fluorescence microscopy using special filter sets. Fluorophores that are most often used in immunohistochemistry are presented in Table 14.1 (see Sect. 14.2).

Fluorophores were introduced to fluorescence microscopy in the early twentieth century, but did not see widespread use until the early 1940s when Albert Coons developed a technique for labeling antibodies with fluorescent dyes, thus giving birth to the field of immunofluorescence (http://www.olympusconfocal.com/theory/fluorophoresintro.html). By attaching different fluorophores to different antibodies, the distribution of two or more antigens can be determined simultaneously in the tissue section and, in contrast to brightfield microscopy, even in the same cells and in the same cell structures (see Chap. 8).

2.5 Colloidal Gold Labels for Electron Microscopy

Colloidal gold is a suspension (or colloid) of sub-micrometer-sized particles of gold in a fluid, usually water. Known since ancient Roman times, it was originally used as a method of staining glass with various shades of yellow, red, or mauve color, depending on the concentration of gold. A colloidal gold conjugate consists of gold particles coated with a selected protein or macromolecule, such as an antibody, protein A or protein G (see Sect. 1.5).

The production of gold-labeled antibodies is not practiced in general histochemical laboratories, but if you have time and inclination to perform colloidal-gold labeling yourself, a good and thorough procedure may be found on the website of Research Diagnostics, Inc. (http://www.researchd.com/gold/gold8.htm). A broad choice of immunogold conjugates is available from Structure Probe, Inc. (http://www.2spi.com/), Aurion (http://www.aurion.nl/), Electron Microscopy Sciences (http://www.emsdiasum.com/microscopy/default.aspx), Energy Beam Sciences, Inc. (http://www.ebsciences.com/company.htm), Dianova (http://www.dianova.de), Nanoprobes, Inc. (http://www.nanoprobes.com/) and some others.

Because of their high electron density, the gold particles are visible in the electron microscope without further treatment. For electron microscopical immunolabeling, the gold particles are manufactured to any chosen size from 6 to 25 nm. The size of the gold particles may be enlarged with subsequent use of silver enhancement technique. The SPI-Mark Silver Enhancement kit (Structure Probe, Inc., http://www.2spi.com/) or AURION R-Gent system from Aurion (http://www.aurion.nl/) provide simple-to-use and sensitive systems for the amplification of immunogold labeling. Silver enhancement occurs during the reduction of silver from one solution (e.g., the Enhancer) by another (e.g., the Initiator) in the presence of gold particles. The reduction reaction causes silver to build up preferentially on the surface of the gold particles, which are conjugated to proteins A and G or to antibodies on the target specimen. Enlarged in this way, the gold particles can be seen at much lower magnifications, even in the light microscope. In the light microscope, the intense brown/black stain produced by silver enhancement of the gold signal gives a greater overall sensitivity and far greater resolution than other immunocytochemical methods, whereby tissues can be counterstained with all the usual staining procedures.

References

Brown JK, Pemberton AD, Wright SH, Miller HRP (2004) Primary antibody–Fab fragment complexes: a flexible alternative to traditional direct and indirect immunolabeling techniques. J Histochem Cytochem 52:1219–1230

Buchwalow IB, Minin EA, Böcker W (2005) A multicolor fluorescence immunostaining technique for simultaneous antigen targeting. Acta Histochemica 107:143–148

Harlow E, Lane D (1999) Using Antibodies: A Laboratory Manual. Cold Spring Harbor Laboratory Press, NY

Lojda Z, Gossrau R, Schiebler T (1976) Enzyme Histochemistry: A Laboratory Manual. Springer, Berlin

Pitt JC, Lindemeier J, Habbes HW, Veh RW (1998) Haptenylation of antibodies during affinity purification: a novel and convenient procedure to obtain labeled antibodies for quantification and double labeling. Histochem Cell Biol 110:311–322

Polak JM, Van Noorden S (1997) Introduction to immunocytochemistry. BIOS Scientific Publishers, Oxford

Sternberger LA (1986) Immunocytochemistry, 3rd Edn. John Wiley and Sons, New York

Van Noorden CJF, Frederiks WM (1993) Enzyme Histochemistry: A Laboratory Manual of Current Methods. Bios Scientific Publishers Ltd., Oxford

Chapter 3
Probes Processing in Immunohistochemistry

The quality of immunohistochemical staining depends on many factors, including fixation, washing and incubation conditions as well as on the appropriate mounting of cell and tissue specimens onto slides. Most of the protocols used in immunohistochemistry are, however, far from being standardized. There are many satisfactory variants of the methods — a collection of useful immunohistochemistry protocols is available on the websites, such as: http://www.ihcworld.com/protocol_database. htm; http://www.histochem.net/; http://www.hoslink.com/immunohis.htm; http:// www.immunoportal.com/. The protocols given in this handbook are practiced in the authors' laboratory.

3.1 Fixation in Immunohistochemistry

An appropriate fixation of biological probes will have a significant impact on the quality of immunohistochemical staining. Specimens subject to immunocytochemical analysis must meet the following technical criteria: (a) cell and tissue preservation should be adequate to characterize the localization of the component of interest, and (b) the antigenicity of the component of interest must still be present and must be accessible to the antibody. To satisfy these demands, fixation should be sufficient to maintain the integrity of the section throughout the whole immunostaining procedure, but not so harsh as to destroy the antigen under study (Hyatt 2002). Maintaining the specimen morphology is the major prerequisite for good immunostaining. If the origin is garbage, do not expect wonderful results. In other words, "Garbage in, garbage out" (Miller 2001).

The choice of fixation is a matter of experience, since the changes in the antigenicity of the component of interest resulting from the treatment with the fixative cannot be predicted. The epitope (amino acid sequence), against which the antibody was raised, may be hidden (masked), modified or completely lost as a result of an inadequate fixation (Shi et al. 2001). No ideal fixative has been found that can be used universally in immunohistochemistry. The choice of the adequate

I.B. Buchwalow and W. Böcker, *Immunohistochemistry: Basics and Methods*,
DOI 10.1007/978-3-642-04609-4_3, © Springer-Verlag Berlin Heidelberg 2010

fixation can be made only by trial. Three main types of fixation in immunocyto-chemistry are:

(a) Air drying
(b) Snap freezing
(c) Chemical fixation

   Air drying, usually with subsequent chemical post-fixation, is applicable for cell smears, cytospins and cryosections. Snap freezing (usually in liquid nitrogen) is routinely employed for tissue probes for subsequent cryosectioning. Chemical fixation is commonly carried out in aldehydes (e.g. formaldehyde) for tissue probes and cultured cell monolayers, and in acetone or alcohols (methanol or ethanol) for cryosections and cell preparations.

3.1.1 Fixation in Alcohols and Acetone

This type of fixation is often used for cryosections, cell smears and cell monolyers, but can not be recommended for fixation of tissue blocks since acetone and alcohols, as opposed to aldehydes, penetrate tissue poorly. Cryosections, cell smears and cell monolyers after short (for 5–15 min) fixation in alcohols or acetone are usually air-dried (for 1 h or overnight), washed in buffered saline and directly subjected to the immunocytochemical analysis.

   However, a limitation of these fixatives is that soft or fatty tissues cannot be adequately stabilized with acetone and alcohols. Moreover, following long immu-nohistochemical staining procedure, acetone-/alcohol-fixed frozen sections may show severe morphological changes. The loss of morphological integrity of the tissue is often associated with a possible loss of the antigen itself. This can be partially prevented by ensuring that tissue cryosections are thoroughly dried both before and after fixation.

   In view of disadvantages associated with the use of cryosections in immunohis-tochemistry, the general trend is that most immunohistochemical investigations both in diagnostic pathology and in basic research studies are carried out on formaldehyde-fixed paraffin-embedded tissue sections.

3.1.2 Fixation in Formaldehyde

Formaldehyde has several advantages over alcohols and acetone, particularly the superior preservation of morphological detail. When the specimen has to be embedded in paraffin or synthetic resin, formaldehyde fixation is the best choice. Formaldehyde is the simplest aldehyde. Its chemical formula is H_2CO. It was first synthesized by the Russian chemist Aleksandr Butlerov in 1859. Discovered to be a tissue fixative originally by the German pathologist Ferdinand Blum in 1893, it is used widely in histology and pathology. Formaldehyde fixative is made up of commercial concentrated formalin (37%–40% solution of formaldehyde) diluted to a 10% solution (3.7%–4% formaldehyde). Formaldehyde dissolved in phosphate buffered saline (PBS, 0.01 M Phosphate buffer, 150 mM NaCl, pH 7.4) can be recommended as a universal fixative for the routine use. Along with PBS, Tris-HCl or cacodylate, buffers with or without isotonic saline may be also used.

Formaldehyde solutions prepared by dissolving and depolymerization of para-formaldehyde (a homopolymer of formaldehyde with empirical formula HO $(CH_2O)nH$, where $n \geq 6$) are free of admixtures of methanol and formic acid. Depolymerized paraformaldehyde is useful in enzyme histochemistry, when the preservation of the enzyme activity is of crucial importance, but it has no advantage over formalin solutions routinely used in pathology and in immunohistochemistry.

In most cases, fixation may be carried out at room temperature. Duration of formalin fixation depends on the nature and the size of the specimen, and may vary from 15 min to 24 h. Longer fixation may be associated with a partial loss of the antigenicity of the component of interest. After formalin fixation, tissue samples are washed in three changes of the buffered saline (PBS) from 15 min to 2 h, but not longer than 24 hours on the whole, since the formaldehyde fixation is partially reversible. After washing in PBS, specimens may be either snap-frozen in liquid nitrogen for subsequent cryosectioning, or dehydrated and embedded in paraffin or synthetic resin.

3.1.3 Effect of Formaldehyde Fixation on Antigen–Antibody Binding

Formaldehyde has several advantages over alcohols and acetone, particularly the superior preservation of morphological detail due to inducing molecular cross-links in proteins. This, however, changes the native three-dimensional protein conformation, thereby altering the normal three-dimensional structure of the epitope (the specific area of the protein where the antibody binds), which makes it more difficult for the antibody to bind to its target. The magnitude of this unfavorable effect varies with the length of fixation, with increasing fixation times causing increasing amounts of antigen damage, also referred to as "antigen masking" (Boenisch 2005). Tissues allowed to sit in formalin over a weekend or longer are subject to increasing amounts of antigen damage. In order to avoid this problem, tissues that will not be processed for a long period of time should be placed in 70% alcohol for long-term storage, since this will minimize adverse effects of antigen damage (Miller 2001). However, if tissue is not overfixed in formalin (which can be considered to be anything over 24 h), formalin is actually an excellent fixative for immunohistochemistry.

3.2 Paraffin Sections for Immunohistochemical Analysis

Formaldehyde fixation and embedding in paraffin enable easy storage and excellent morphological detail and resolution. Routine formaldehyde fixation and the rest of the procedure for paraffin embedding may alter the conformation of protein macromolecules, negatively affecting antigen–antibody interaction and consequently decreasing the intensity of the final reaction in the immunohistochemical procedure. However, these deleterious effects can in many instances be easily overcome by employing appropriate epitope retrieval techniques such as heat-induced antigen retrieval (Boenisch 2005; Shi et al. 1991, 2001) (see Sect. 6.1.1). Development of the heat-induced antigen retrieval technique, a simple method of boiling paraffin-embedded tissue sections in buffered water solutions, rendered routine formalin-fixed, paraffin-embedded tissues more attractive for immunohistochemistry than cryosections. Cryosections may be used if no paraffinized tissue is available or if the antigen cannot be detected after formalin fixation and paraffin-embedding, due to protein cross-linking.

3.2.1 Embedding and Cutting

Aldehyde fixation usually has to be followed by dehydration, clearing, embedding (commonly in paraffin) and cutting the specimen on the microtome. Dehydration is the first step in the processing of fixed tissues. The water in the specimen should be replaced, first with alcohol or acetone and then with a paraffin solvent (clearing agent) such as xylene or toluene. After fixation, dehydration and clearing, tissue blocks are impregnated by paraffin wax.

Paraffin embedding schedule
- Cut fixed tissues into blocks up to 5–10 mm thick.
- Put the pieces of tissue into embedding cassettes.
- 70% ethanol, two changes for 1 h at room temperature.
- 95% ethanol, two changes for 1 h at room temperature.
- 100% ethanol, two changes for 1 h at room temperature.
- Xylene or toluene, two changes for 1 h at room temperature.
- Paraffin wax, two changes for 1 h in a 58°C–64°C oven. The second change may be overnight.
- Embedding tissues into paraffin blocks.

The whole tissue processing from fixation to embedding in paraffin can be performed manually or automated by means of processing machines. Cutting of paraffin-embedded tissues is performed by means of microtomes. Trimmed paraffin blocks are cut at 3–10 μm (5 μm is commonly used).

3.2.2 Mounting Paraffin Sections onto Slides

Prior to immunohistochemical staining, paraffin sections must be properly mounted onto slides, and then deparaffinized and rehydrated. To help adherence to the glass and decrease the chances of sections dissociating from the slides, paraffin tissue sections should be mounted on tissue-adhesive-coated slides. The use of tissue-adhesive-coated slides is especially important for paraffin tissue sections undergoing heat-induced antigen retrieval (see Sect. 6.1.1).

Adhesive slides are available from numerous manufacturers, such as Polysine microslides (Menzel Gläser, Braunschweig, Germany) or Super Frost Plus slides (Fisher Scientific), although Miller (2001) reported that he had best results by preparing his own silanized slides, based on a method published by Henderson (1989). In difficult cases, he had good results by taking routine silanized slides and "double dipping" them a second time in the silane adhesive solution. These "double dipped" silanized slides were far more effective than any of the numerous commercially-prepared adhesive slides.

Preparation of APES (amino-propyl-tri-ethoxy-silane)-treated slides
– Wash glass slides in detergent for 30 min.
– Wash glass slides in running tap water for 30 min.
– Wash glass slides in distilled water 2 × 5 min.
– Wash glass slides in 95% alcohol 2 × 5 min.
– Air-dry for 10 min.
– Immerse slides in a freshly prepared 2% solution of APES in dry acetone for 5 s.
– Wash briefly in distilled water twice.
– Dry overnight at 42°C and store at room temperature.

Note: 300 ml of APES solution is sufficient to do 200 slides. Treated slides can be kept indefinitely.

When mounting paraffin sections onto slides, it is advisable to use deionized water in the water bath in order to avoid sectioning artifacts. When mounted onto slides, tissue paraffin sections must be properly dried. Drying increases adhesion of the tissue sections to the surface of the glass slide. For the drying of paraffin sections mounted on adhesive-covered slides, a number of options are available depending upon how much time one desires to devote to this step (Miller 2001). Drying the slides in a 56°C oven for 1 h works well, and if you have the luxury of drying your slides at 37°C overnight or longer up to 3 days, this is even better. Mounted paraffin tissue sections must be deparaffinized and rehydrated before use, to ensure that the antibodies have full access to the tissue antigen. The standard procedure for deparaffinization is described below.

Procedure for deparaffinization and rehydration

(1) Dry tissue paraffin sections at 56°C for 1 h or longer up to 3 days.
(2) Place the slides in a cuvette containing sufficient xylene to cover the tissue completely, and incubate for 5 min with gentle shaking.

(3) Transfer the slides to a cuvette containing fresh xylene and repeat step (2) for a further wash lasting again 5 min. Important: use fresh xylene for each wash.
(4) Wash the slides twice in 100% ethanol (for 3 min each time).
(5) Wash the slides in 90–96% ethanol for 3 min.
(6) Wash the slides in 70% ethanol for 3 min.
(7) Rinse the slides in distilled water.
(8) Follow procedure for antigen retrieval if required.
(9) Rinse the slides in a buffer of choice.
(10) Circle sections with a hydrophobic barrier pen (e.g., Dako Pen, #S2002). Do not allow sections to dry for the remaining procedure.
(11) Proceed with immunohistochemical staining protocol.

After deparaffinization and rehydration of sections, you may continue with the antigen retrieval if affordable (see Sect. 6.1.1) and the immunohistochemical staining protocol.

3.3 Cryosections for Immunohistochemical Analysis

For cryosectioning, tissue samples are quickly frozen with or without freeze-embedding medium (e.g., Tissue Tek; Miles Laboratories) and stored at –80°C until analysed. Optionally, aldehyde prefixation can also be used for tissue and organ probes before snap-freezing. Cutting of frozen tissue blocks is performed with a cryostat (a microtome mounted in a freezing cabinet).

Preparation of cryosections for immunohistochemical analysis

(1) Small pieces of tissue (approx. 5 mm) are quickly frozen with or without freeze-embedding medium (e.g.,Tissue Tek; Miles Laboratories) and stored at –80°C until analysed.
(2) Cut frozen sections at 5–10 μm. Pick up on adhesive-coated slides.
(3) Air dry sections overnight or at least 1 h at room temperature.
(4) Fix sections in acetone or methanol (room temperature) for 10–15 min.
(5) Air dry sections.
(6) Circle sections with a hydrophobic barrier pen (e.g., Dako Pen, #S2002) and proceed with immunohistochemical staining protocol. Do not allow sections to dry for the remaining procedure.
(7) Alternatively, skip step (6) and store sections at −20°C until needed.

This protocol will generally produce good quality immunohistochemical staining. Frozen tissue sections normally do not need a heat-induced antigen retrieval step.

As for paraffin sections, it is advisable to mount cryosections also onto adhesive-coated slides in order to decrease the chances of sections dissociating from the slides in the course of immunohistochemical staining. Once mounted on slides, cryosections are air-dried and fixed, usually in acetone or methanol. Aldehyde fixation is also possible. After washing in a buffer, cryosections are ready for immunostaining, but they can also be stored in PBS in the refrigerator until they can be stained. Miller (2001) found that B and T cell antigens survive well at least for 5 days in PBS in the refrigerator, and other antigens (cytokeratin, etc.) survive for months. Morphology is also well-preserved for months (up to 16 months, provided the PBS does not get contaminated by bacterial growth).

For longer storage, air-dried cryosections are best kept at $-20°C$ or at $-80°C$ until needed. When required, allow the slides to warm at room temperature for 5 min, then fix again in acetone (optionally) for 5 min and rinse in a washing buffer of choice.

3.4 Buffers for Washing and Antibody Dilution

The type of buffer selected for washing and for dilution of antibodies will have a significant impact on the quality of immunohistochemical staining. In immunohistochemistry, buffers near physiologic pH are traditionally used. Because this pH lies within the range of isoelectric points of immunoglobulins, the optimal immunoreactivity of antibodies is generally awaited at pH 7.4 (Boenisch 1999). Immunohistochemical staining is usually carried out with antibodies dissolved in phosphate buffered saline (PBS) or in Tris-buffered saline (TBS). There exists no convincing evidence concerning the preference of either PBS or TBS for diluting antibodies. As an antibody diluent, we use in our laboratory routinely PBS. However, phosphate-containing buffers can not be used when working with alkaline phosphatase-conjugated antibodies. In such cases, TBS is the only alternative. Only freshly prepared buffers should be used. Bacterial contamination, which can occur in buffers stored at room temperature, may affect the quality of the staining.

10x PBS (1M PBS, pH 7.4):

Na$_2$HPO$_4$ (anhydrous)	10.9 g
NaH$_2$PO$_4$ (anhydrous)	3.2 g
NaCl	90 g
Distilled water	1,000 ml

Mix to dissolve and adjust pH to 7.4

Store this solution at room temperature. Dilute 1:10 with distilled water to obtain a 100 mM working solution before use and adjust pH if necessary.

Note: Do not use PBS for alkaline phosphatase-conjugated antibodies, since phosphate is an inhibitor of alkaline phosphatase.

10x TBS (1M TBS, pH 7.4)
– Dissolve 121 g Tris Base and 90 g NaCl in 500 ml of distilled water
– Adjust pH to 7.4 with approximately 200–300 ml 2 M HCl.

- Q.S. to 1 liter with distilled water
- Store this solution at room temperature. Dilute 1:10 with distilled water to obtain a 100 mM working solution before use, and adjust pH if necessary.

It has also been proposed that new monoclonal antibodies be first examined by immunohistochemistry in several dilutions of twofold increments beyond those generally recommended by vendors, using 0.05 M Tris buffers of pH 6.0 and 8.6 without NaCl. Thereby, many monoclonal antibodies could be used in dilutions up to eightfold higher than those recommended by the vendor (Boenisch 1999). To the time of this writing, we have not yet had the opportunity to evaluate this apparently interesting approach.

Including a small amount of some type of antimicrobial agent (0.05% sodium azide, 0.05% thimerosal or similar agent) in order to prevent microbial contamination is only advisable for longer storage of the antibody solution. Generally, antibody solutions are, however, prepared ex tempore. 1% BSA has been claimed to stabilize diluted antibody solutions (Ciocca et al. 1983). However, this may make sense only if you intend to store diluted antibody solutions for many weeks. Moreover, some antibodies can bind BSA or other serum containing reagents, which may results in reducing antibody affinity. Do not use serum-containing solution to dilute avidin/streptavidin conjugates, since serum may contain biotin, therefore reducing reaction activity for avidin/streptavidin conjugates.

Adding a small amount (0.1%) of detergents such as Tween 20 to the buffer assists in uniform spreading of the reagents, due to reducing surface tension in washing solutions, and allows the slides to remain wet longer. There is, however, no role for detergent in permeabilization of fixed biological probes. Once tissues or cells are fixed (formalin, acetone, ethanol, freezing, drying, etc.), they no longer can select intracellular traffic of molecules, and antibodies can easily cross cell membranes.

10x PBS-Tween 20 (0.1M PBS, 0.5% Tween 20, pH 7.4)

Na_2HPO_4	10.9 g
NaH_2PO_4	3.2 g
NaCl	90 g
Distilled water	1,000 ml

Mix to dissolve and adjust pH to 7.4 and then add 5 ml of Tween 20.

Store this solution at room temperature. Dilute 1:10 with distilled water to obtain a 100 mM working solution before use and adjust pH if necessary.

Thorough and effective rinsing is very important between the various steps of the procedure. Most laboratories, as well as our laboratory, employ either PBS or TBS as a rinse buffer. Adding Tween 20 to the rinse buffer should aid in effective rinsing. Referring to Dave Tacha (BioCare, Walnut Creek, CA, USA), Miller (2001) recommends the use of distilled water with Tween 20 (2 ml Tween 20 in 10 l of distilled water) rather than PBS or TBS for the rinsing steps. To the time of this writing, we have not yet had the opportunity to evaluate this interesting innovation, which apparently might save substantial amounts of reagent cost as well as reagent preparation time.

3.5 Mounting Following Immunohistochemical Staining

Mounting of the specimen under the coverslip is the last step in the immunohisto-chemical staining protocol (see Sect. 4). Mounting medium serves to help the coverslip to adhere to the slide bearing the tissue section or cytological preparation, protects the specimen and the immunohistochemical staining from physical damage, and improves the clarity and contrast of the image during microscopy. The choice of the mounting medium depends on the label used to visualize the antigen. Either aqueous or non-aqueous mounting media is suitable, depending on the chromogenic substrate used to detect the antigen.

In brightfield microscopy, aqueous mounting media are generally suitable for all enzyme chromogenic labels. Organic mounting media, such as Canada balsam or DPX (see Table 3.1) are only used for enzyme chromogenic labels that are not soluble in the alcohols during dehydration (e.g., a peroxidase substrate DAB), although the faster procedure using an aqueous medium can also be used. Another chromogenic label for peroxidase, AEC, is soluble in alcohol and clearing agents, and therefore must be mounted in aqueous mounting media. Most alkaline phosphatase substrates, such as Vector® Red, Vector® Black, and BCIP/NBT, can be permanently mounted in non-aqueous media; however, aqueous medium can also be used. Always observe the guidelines given on reagent product inserts (see Sect. 2.3). For fluorescence microscopy, aqueous mounting media must contain antifade agents (free-radical scavengers) preventing photobleaching of fluorophore labels under exposure to excitation light.

Commercial ready-to-use mounting media may be excellent for what you want to do; however, they are not rather cheap, and in most cases their formulation is a trade secret. Media made in the laboratory are cheaper than commercial aqueous

Table 3.1 Most commonly used mounting media (adapted from Renshaw (2007))

Aqueous mounting media for brightfield microscopy	Aqueous mounting media for fluorescence microscopy	Organic mounting media
Ready-to-use adhesive aqueous mounting media are available from Immunotech, Dako, Vector Laboratories, Molecular Probes, etc.	80–90% glycerol in PBS containing an antifading agent such as 0.1% p-phenylendiamine or 2% n-propylgallate: (nonadhesive)	Canada balsam: (sets hard)
Glycerin jelly (glycerol + gelatin): (adhesive)	VectaShield from Vector Laboratories, Burlingame, USA: (nonadhesive)	Dibutyl phthalate xylen (DPX), also known as Distrene plasticiser: (sets hard)
50% (v/v) glycerol in PBS: (nonadhesive)	ProLong B from Molecular Probes: (sets hard)	–

mountants, and do not contain any secret ingredients. If you have the inclination and the time to prepare aqueous mounting media yourself, good and thorough procedures may be found on the website: http://ihcworld.com/_protocols/histology/aqueous_mounting_medium.htm.

Procedure for non-aqueous mounting

– Transfer slides to a buffer of choice and agitate gently for a few seconds. If you performed nuclear counterstaining with hematoxylin, transfer to tap water for 5–10 min.
– Transfer slides to 70% ethanol and agitate gently for a few seconds.
– Transfer slides to 100% ethanol for 3 min.
– Transfer slides to xylene for 3 min.
– To each slide, apply one or two drops of non-aqueous mounting medium to the uncovered tissue and apply the coverslip. Apply the coverslip at an inclined angle to the slide, and lower it gently to avoid trapping air bubbles.
– Leave the slides to air-dry overnight, or dry in an incubator at 56–60°C for 15 min.

Procedure for aqueous mounting

– Transfer slides to a buffer of choice and agitate gently for a few seconds. If you performed nuclear counterstaining with hematoxylin, transfer to tap water for 5–10 min.
– Add one or two drops of aqueous mounting medium to the stained tissue and apply the coverslip.
– Apply the coverslip at an inclined angle to the slide and lower it gently to avoid trapping air bubbles.

3.6 Storage Following Immunohistochemical Staining

Organic mounting media like Canada balsam or DPX will dry quickly without the coverslip lifting and without adversely affecting the specimen, thus being ideal for storage of such preparations at room temperature for many years on the bench. However, most enzymatic chromogen labels will readily fade if exposed to light for extended periods. Storage at +4°C in the dark is only required for specimens mounted with aqueous media to reduce the rate of evaporation, causing the coverslip to lift and to prevent photobleaching of fluorophore labels (Renshaw 2007).

References

Boenisch T (1999) Diluent buffer ions and pH: their influence on the performance of monoclonal antibodies in immunohistochemistry. Appl Immunohistochem Mol Morphol 7:300–306
Boenisch T (2005) Effect of heat-induced antigen retrieval following inconsistent formalin fixation. Appl Immunohistochem Mol Morphol 13:283–286

Ciocca DR, Adams DJ, Bjercke RJ, Sledge GW, Edwards DP, Chamness GC, Mcguire WL (1983) Monoclonal-antibody storage-conditions, and concentration effects on immunohistochemical specificity. J Histochem Cytochem 31:691–696
Hyatt MA (2002) Microscopy, Immunohistochemistry, and Antigen Retrieval Methods: For Light and Electron Microscopy. Plenum, New York
Henderson C (1989) Aminoalkylsilane: an inexpensive, simple preparation for slide adhesion. J Histochem Cytochem 12(2):123–124
Miller RT (2001) Technical immunohistochemistry: achieving reliability and reproducibility of immunostains. Society for Applied Immunohistochemistry, 2001 Annual Meeting. http://www.ihcworld.com/_books/Technical-IHC.pdf
Renshaw S (2007) Immunohistochemistry. Scion Publishing, Cambridge
Shi SR, Key ME, Kalra KL (1991) Antigen retrieval in formalin-fixed, paraffin-embedded tissues: an enhancement method for immunohistochemical staining based on microwave oven heating of tissue sections. J Histochem Cytochem 39:741–748
Shi SR, Cote RJ, Taylor CR (2001) Antigen retrieval techniques: current perspectives. J Histochem Cytochem 49:931–938

Chapter 4
Working with Antibodies

Deparaffinized and rehydrated tissue sections and cytological preparations can be subjected to antigen retrieval (see Sect. 6.1.1) and to blocking steps (see Chap. 5) and finally put through the rest of the immunohistochemical protocol. The core of all immunohistochemical protocols is the antigen–antibody binding. Binding of the primary antibody to its antigen is the key step responsible for good quality immunohistochemical staining. Upon binding the primary antibody to the antigen, you may detect the bound primary antibodies in your specimen using a labeled secondary antibody.

Correspondingly, two principal immunohistochemistry methods are employed — direct and indirect. The direct method is a straightforward one-step process that creates a direct reaction between the antigen and the labeled antibody. The indirect method requires the use of two antibodies — a primary unlabeled antibody and a secondary labeled antibody.

4.1   Direct Immunostaining Method

The direct method (one antibody layer) involves an antibody reacting directly with the antigen in tissue sections. In this method (Fig. 4.1), the primary antibody must be labeled. Common labels include fluorescent dye, biotin or enzymes. (see Chap. 2). The fluorophore label can be visualized directly using fluorescent microscopy. The biotin label can be detected using streptavidin conjugated with a fluorophore or an enzyme; the latter must be visualized through an enzyme chromogenic system (see Sect. 2.3). See "Basic protocol for direct immunolabeling" below.

Basic protocol for direct immunostaining
  1. Deparaffinize and rehydrate tissue sections. Rinse in distilled water for 5 min.
  2. Antigen retrieval: place sections in a Coplin jar with antigen retrieval solution of choice (e.g. 10 mM citrate acid, pH 6) and heat at 90°C–110°C (depending

I.B. Buchwalow and W. Böcker, *Immunohistochemistry: Basics and Methods*,
DOI 10.1007/978-3-642-04609-4_4, © Springer-Verlag Berlin Heidelberg 2010

Fig. 4.1 Direct
immunostaining method

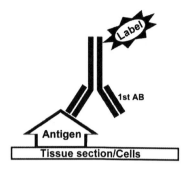

on tissue) in a microwave, steamer, domestic pressure cooker or autoclave. See
Sect. 6.1.1.

3. Washes and dilutions: wash sections in 10 mM sodium phosphate buffer,
   pH 7.5, 150 mM NaCl (PBS) for 2 × 3 min*. PBS is used for all washes
   and dilutions. Other buffers such as Tris buffered saline (TBS) may also be
   used.
4. Circle sections with a hydrophobic barrier pen (e.g., Dako Pen, #S2002). Do
   not allow sections to dry for the remaining procedure.
5. Blocking step: incubate sections for 20 min with normal serum blocking
   solution (see Sect. 5.1).
6. Primary antibodies: blot excess blocking solution from sections and incubate
   for 60 min at room temperature or overnight at + 4°C with a correspondingly
   diluted primary antibody labeled with a fluorophore or biotin. Wash sections in
   PBS for 2 × 3 min.
7. Make it visible: the fluorophore label can be visualized directly using fluorescent microscopy. The biotin label (see Sect. 6.2.1) can be detected using
   streptavidin conjugated with an enzyme; the latter must be visualized through
   an enzyme chromogenic system. Incubate sections with an appropriate enzyme
   substrate until optimal color develops (see Sect. 2.3).
8. Counterstaining: counterstain nuclei if necessary, e.g., with DAPI** for
   fluorescence microscopy or with hematoxylin (see Sect. 7.4) for brightfield
   microscopy.
9. Optional: when using a fluorescent label, a short treatment (1–3 min) with 4%
   formaldehyde in PBS before mounting in water-soluble media is recommended
   for blocking the detachment of the fluorophore from the antibody; this preserves the staining pattern for a longer storage. Wash sections in PBS for
   2 × 3 min.
10. Mounting: mount sections in aqueous medium or balsam for brightfield micros-
    copy or in anti-fade medium for fluorescence microscopy (see Sect. 3.2.2).
    Notes: *All incubations are at room temperature unless otherwise noted.
    **Nuclear dyes (DAPI, Hoechst 33342 and Propidium Iodide) supplied as
    lyophilized solids are usually reconstituted in methanol. The stock solutions
    (5 mg/ml) are stable for many years when stored frozen at $\leq -20°C$ and

protected from light. Before use, the stock solution is further diluted in PBS to the final concentration of 5 μg/ml.

The direct method is applicable when using antibodies with high avidity and for localization of high density antigens (>10 K molecules/cell). To localize antigens expressed at a lower level, the indirect method (two-antibodies layer, see Fig. 4.2) is affordable. On the other hand, the direct method offers significant benefits in techniques compared with indirect immunostaining. Without the problems of crossover and/or nonspecific binding from secondary reagents it is far easier to obtain good quality data. A reduction in the number of incubation and wash steps also makes assays less tedious and saves time. Unfortunately, labeled primary antibodies are not always available commercially, and it might be necessary to make custom-labeled antibody against the antigen of your interest. Modern developments in antibody labeling make in-house antibody labeling possible (see Chap. 2). If, however, you are not inclined to label your antibody yourself but instead prefer to invest your money rather than your time, check the websites for "Custom Labeling Service", such as http://www.biocompare.com/ProductListings/16861/ Antibody-Labeling-Services.html, or http://www.innovabiosciences.com/services/ custom_labeling.php.

4.2   Indirect Immunostaining Method

Due to its higher sensitivity, the indirect method is in more common use than direct methods. In indirect immunostaining, the bound unlabeled primary antibody (first layer) is visualized with a secondary antibody (second layer) bearing label, such as a fluorophore, biotin or an enzyme (Fig. 4.2). See "Basic protocol for indirect immunolabeling" below. This method is more sensitive as a result of signal amplification through several secondary antibodies binding to different antigenic sites on the primary antibody on both Fc and Fab fragments.

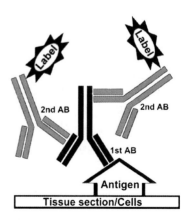

Fig. 4.2 Indirect immunostaining method

Basic protocol for indirect immunostaining

1. Deparaffinize and rehydrate tissue sections. Rinse in distilled water for 5 min.
2. Antigen retrieval: place sections in a Coplin jar with antigen retrieval solution of choice (e.g. 10 mM citrate acid, pH 6) and heat at 90°C–110°C (depending on tissue) in a microwave, steamer, domestic pressure cooker or autoclave. See Sect. 6.1.1.
3. Washes and dilutions: wash sections in 10 mM sodium phosphate buffer, pH 7.5, 150 mM NaCl (PBS) for 2×3 min*. PBS is used for all washes and dilutions. Other buffers such as Tris buffered saline (TBS) may also be used.
4. Circle sections with a hydrophobic barrier pen (e.g., Dako Pen, #S2002). Do not allow sections to dry for the remaining procedure.
5. Blocking step: incubate sections for 20 min with PBS containing 5% normal serum of species in which the secondary antibodies were raised.
6. Primary antibodies: blot excess blocking solution from sections and incubate for 60 min at room temperature or overnight at +4°C with a correspondingly diluted unlabeled primary antibody. Wash sections in PBS for 2×3 min.
7. Secondary antibodies: incubate sections for 30–60 min at room temperature with a labeled secondary antibody raised against the corresponding IgG of the primary antibody. Wash sections in PBS for 2×3 min.
8. Make it visible: the fluorophore label can be visualized directly using fluorescent microscopy. The enzyme label must be visualized through an enzyme chromogenic system. Incubate sections with an appropriate enzyme substrate until optimal color develops (see Sect. 2.3). The biotin label can be detected using streptavidin conjugated with an enzyme or fluorophore (see Sect. 6.2.1).
9. Counterstaining: counterstain nuclei if necessary, e.g., with DAPI** for fluorescence microscopy or with hematoxylin (see Sect. 7.4) for brightfield microscopy.
10. Optional: when using a fluorescent label, a short treatment (1–3 min) with 4% formaldehyde in PBS before mounting in water-soluble media is recommended for blocking the detachment of the fluorophore from the antibody; this preserves the staining pattern for a longer storage. Wash sections in PBS for 2×3 min.
11. Mounting: mount sections in aqueous medium or balsam for brightfield microscopy or in antifade medium for fluorescence microscopy (see Sect. 3.2.2).
    Notes: *All incubations are at room temperature unless otherwise noted.
    **Nuclear dyes (DAPI, Hoechst 33342 and Propidium Iodide) supplied as lyophilized solids are usually reconstituted in methanol. The stock solutions (5 mg/ml) are stable for many years when stored frozen at $\leq$–20°C and protected from light. Before use, the stock solution is further diluted in PBS to the final concentration of 5 µg/ml.

The addition of the secondary antibody step allows for signal amplification of the process. However, for low density antigens ($>$2 K, $<$ 10 K molecules/cell) it is recommended to employ further amplification of the immunocytochemical signal with tyramide, with ABC method or with polymer-conjugated technology such as "EnVision System" developed by DakoCytomation (see Sect. 6.2).

4.3 The Choice of Antibodies

A vast amount of information on both monoclonal and polyclonal antibodies can be retrieved using Internet search engines (http://www.antibodies-online.com/; http://www.linscottsdirectory.com/search/antibodies) as well as Web portals provided by antibody manufacturers (http://www.abcam.com/index.html; http://www. biozol.de/2008/; http://www.dianova.de/produkte/p_recherche.php; http://www. biocompare.com/ProductCategories/2045/Antibody-Search.html; http://probes. invitrogen.com/handbook/). Antibody Resource Page (http://www.antibodyresource. com/) is a guide designed by scientists for scientists to find companies that sell catalog antibodies and custom monoclonal and polyclonal antibodies. Antibody lists on these websites can be browsed by antigen, conjugate, research area, application, host, target species (reactivity), clone, etc. Many of these websites provide additional URL links to other resources including text books, bibliographies, journal articles, meeting and committee proceedings, suppliers of antibody reagents and services, and applications for use of antibodies.

The decision regarding whether to use a polyclonal antibody or monoclonal antibody depends on a number of factors, the most important of which are its intended use and whether the antibody is readily available from commercial suppliers or researchers. The concentration and purity levels of specific antibody are higher in monoclonal antibodies. However, polyclonal antibodies are known to have a higher affinity and a better specificity than monoclonal antibodies, because they are produced by a large number of B cell clones each generating antibodies to a specific epitope, and because polyclonal sera are a composite of antibodies with unique specificities (Lipman et al. 2005).

4.3.1 The Choice of Primary Antibodies

Purified antibodies are usually prepared from whole serum, culture supernatants or ascites by affinity chromatography (see Table 4.1). Monoclonal antibodies have traditionally been produced in a mouse and to a lesser extent in a rat. Cloning techniques produce highly specific monoclonal antibodies. However, polyclonal antibodies produced by rabbits are known to have a higher affinity than antibodies raised in mice, which means — to put it simply — a better antigen recognition. Additionally, the rabbit immuno-response recognizes antigens (epitopes) that are not immunogenic in mice. It is often for this reason that rabbit polyclonal antibodies are developed and used in many research, and even some diagnostic, applications. In order to combine benefits of better antigen recognition by polyclonal antibodies with the specificity and consistency of a monoclonal antibody, Epitomics (http:// www.epitomics.com/) developed a proprietary unique method for making mono-clonal antibodies from rabbits.

If possible, your primary antibody should not be raised in the same species as the tissue under study, since antigen detection with the use of antibodies on

Table 4.1 Quantity of specific antibodies from various sources (adapted from the Roche Molecular Biochemicals pamphlet "Lab FAQS")

Source	Type	Quantity of total antibody	Quantity of specific antibody	Contaminating antibodies	Purity of specific antibody
Serum	Polyclonal	10 mg/ml	<1 mg/ml	Other serum antibodies	10% (except for antigen affinity column)
Culture supernatants with 10% FCS	Monoclonal	1 mg/ml	0.01–0.05 mg/ml	Background from calf serum	>95%, no cross reaction, high quality
Culture supernatant with serum-free medium	Monoclonal	0.05 mg/ml	0.05 mg/ml	None	>95%, no cross reaction, high quality
Ascites	Monoclonal	1–10 mg/ml	0.9–9 mg/ml	Background from mouse antibodies	Max. 90%, cross reactions possible

homologous tissues (e.g., mouse antibody on mouse tissue) is complicated by severe background staining due to the fact that secondary antibody binds to primary antibody as well as to irrelevant endogenous tissue immunoglobulins. If the wanted antibody from another species is not available, you have to preincubate your specimen with unconjugated monovalent Fab fragments in order to block endogenous tissue immunoglobulins that are responsible for severe background staining (see Chap. 9). Alternatively, you may haptenylate your primary antibody. Haptenylated primary antibody can be visualized using secondary antibody recognizing the corresponding hapten. This makes it possible to evade application of the secondary antibody that can bind both to the primary antibody and to irrelevant endogenous tissue immunoglobulins homologous to the primary antibody (see Sect. 2.2).

4.3.2 The Choice of Secondary Antibodies

Secondary antibodies are offered in three different forms: whole IgG, F(ab')$_2$ fragments, and Fab fragments. The whole IgG form of antibodies is suitable for the majority of immunodetection procedures, and is the most cost-effective.

F(ab')$_2$ fragments of antibodies are generated by pepsin digestion of whole IgG antibodies to remove a portion of the Fc region (see Sect. 1.1). Elimination of nonspecific binding of the Fc region of antibodies to Fc receptors on cell surfaces in immunocytochemical applications leads to lower background, better signal-to-noise ratios, and increased sensitivity. Moreover, F(ab')2 fragments diffuse better through tissue and through cell membranes than intact IgG because they are smaller, leading to faster and more efficient staining. F(ab')2 fragments are used for specific applications, such as to avoid binding of antibodies to Fc receptors on cell surfaces (see Sect. 5.1) or binding to Protein A or Protein G (see Sect. 1.5).

Fab fragments of antibodies are generated by papain digestion of whole IgG antibodies to remove the entire Fc portion. These antibodies are monovalent, containing only a single antigen binding site. They can be used for labeling of primary antibodies (see Sect. 2.2) and for blocking endogenous tissue immunoglobulins, when working with antibodies on homologous tissues (e.g., mouse antibody on mouse tissue) (see Chap. 9).

Caution should be exercised, since secondary antibodies against immunoglobulins from one species may cross-react with a number of other species unless they have been specifically preabsorbed against the cross-reacting species. Antibodies that have been preabsorbed against other species are recommended when the possible presence of immunoglobulins from other species may lead to interfering cross-reactivities. For example, only use antimouse IgG preabsorbed against rat IgG to detect a mouse primary antibody in rat tissue which contains endogenous rat immunoglobulins. For more information helpful in the choice of antibodies, the reader is referred to the website of Jackson ImmunoResearch Laboratories Inc. (http://www.jacksonimmuno.com/technical/select.asp).

4.4 Optimal Concentration of the Antibody

After having purchased a new antibody, it is necessary to determine its optimal working concentration before putting it into the experiment. This is a task that confronts all immunohistochemists, often being the main reason for frustrating results (or perpetual mediocrity) experienced by many laboratories. This predicament is described in more detail by Miller (2001).

The optimal working concentration of the primary antibody is a function of many factors, such as antigen density, loss of antigen activity in the course of specimen preparation, antibody avidity, and sensitivity of the immunostaining procedure. For the vast majority of polyclonal and monoclonal primary antibodies in standard immunohistochemical procedures, their working concentration lies within the range of 1–5 µg/ml. This value can be substantially reduced when using antigen retrieval (see Sect. 6.1) and/or signal amplification (see Sect. 6.2). Secondary antibodies are usually applied at the working concentration approximately 5–10 µg/ml.

Purified antibodies are usually prepared from whole serum by affinity chromatography. Concentration of affinity-purified specific antibodies is normally specified by the manufacturer in the antibody data sheet. When starting with a new antibody of unknown concentration or with a serum, an average quantity of specific antibodies from various sources can be assessed from Table 4.1.

The optimal working concentration can be determined in a series of stains with the antibody in several dilutions of twofold increments (*e.g.*, 1:25, 1:50, 1:100, 1:200, 1:400, 1:800, 1:1600). The resulting antibody concentration can be calculated according to the following formula.

Example:

(**A**) Final volume* (e.g. 1,000 μl)
(**B**) Desired antibody concentration (e.g. 2 μg/ml)
(**C**) Antibody concentration listed by manufacturer (e.g. 100 μg/ml)
(**X**) Volume of antibody required (X ml)
$\mathbf{X} = (\mathbf{A} \times \mathbf{B}) : \mathbf{C} = 20$ μl

Note: *Final volume may be approximated as the volume of the antibody diluent in case of high dilutions. For low dilutions (les than 1:10), the final volume is calculated as follows: final volume = volume of antibody required + volume of antibody diluent.

The "best" stain is then the optimal for that particular primary antibody if you do not observe any nonspecific background staining in control incubations (see below).

4.5   Specificity Controls in Immunohistochemistry

Immunohistochemistry is a powerful method for identification of proteins in cells and tissues, but control procedures are necessary. The specificity of the results depends on two independent criteria: the specificity of the antibody and of the method used (Burry 2000). The antibody specificity is best determined by immunoblot and/or immunoprecipitation. The specificity of the method is best determined by both a negative control, replacing the primary antibody with the corresponding IgG (must be the same species as primary antibody at the same concentration), and a positive control, testing the primary antibody with tissues known to contain the antigen.

An additional control for the antibody specificity is the so-called absorption or preabsorption control, in which the primary antibody (prior to its use) is incubated for 1 h with a tenfold molar excess of the purified antigen. Absent or greatly diminished immunostaining should be obtained after application of this preabsorbed antibody. However, it is sometimes difficult to obtain the purified antigen; therefore, it is rarely used routinely in immunohistochemical staining. Moreover, absorption of the antibody with the purified antigen does not always indicate that the antibody has bound to the same protein in the tissue (Burry 2000).

When fluorescent dyes are used in the experiments, autofluorescence (or natural fluorescence) of some tissue components can cause background problems and complicate the use of fluorescence microscopy. The simplest test is to mount sections in PBS or in any water-soluble medium and to view the specimen with a fluorescence microscope. If autofluorescence is detected in the tissue sections, one of the solutions is to avoid use of fluorescent method and to choose enzyme or other labeling methods. A more intelligent way, however, is to use a wavelength for excitation that is out of the range of natural autofluorescence. The longer wavelength excitation of fluorophores (such as Alexa Fluor 647, Cy5 or Cy7) in infrared spectrum region can be chosen to avoid autofluorescence which is annoying only at shorter wavelengths. The reader will find more in Chap. 5 about how to outsmart autofluorescence.

References

Burry RW (2000) Specificity controls for immunocytochemical methods. J Histochem Cytochem 48:163–166

Lipman NS, Jackson LR, Trudel LJ, Weis-Garcia F (2005) Monoclonal versus polyclonal antibodies: distinguishing characteristics, applications, and information resources. ILAR J 46:258–268

Miller RT (2001) Technical immunohistochemistry: achieving reliability and reproducibility of immunostains. Society for Applied Immunohistochemistry, 2001 Annual Meeting. http://www.ihcworld.com/_books/Technical-IHC.pdf

Chapter 5
Background Staining, Autofluorescence and Blocking Steps

Background staining has increased the amount of gray hair on the heads of histochemists. In most cases, background staining is not caused by a single factor. Along with Fc receptors, frequent causes of background staining are endogenous enzyme activity, if you use peroxidase or alkaline phosphatase as enzyme markers on your secondary antibodies, and endogenous biotin when using a streptavidin or avidin reagents (http://www.ihcworld.com/; http://www.protocol-online.org/prot/ Immunology/). When fluorescent dyes are used in experiments, autofluorescence (or natural fluorescence) of some tissue components can cause background problems and complicate the use of fluorescence microscopy.

5.1 Fc Receptors

Fc receptors present on the cell membrane of macrophages, monocytes, granulocytes, lymphocytes, and some other cells may, in theory, non-specifically bind Fc portion of antibodies used for immunolabeling. Typically, blocking endogenous Fc receptors involves using normal serum derived from the same species used to produce the secondary antibody. To block endogenous Fc receptors, incubate sections for 15–30 min with normal serum (2–5% v/v) from the host species of the secondary antibody.

Normal Serum Blocking Solution
– 2–5% normal serum (blocking)
– 0.05% Tween 20 (detergent and surface tension reducer; optional)
– 0.05% sodium azide* (preservative, if you intend to store this solution for a long time)
– 0.01M PBS, pH 7.4
– Mix well and store at 4°C

I.B. Buchwalow and W. Böcker, *Immunohistochemistry: Basics and Methods*,
DOI 10.1007/978-3-642-04609-4_5, © Springer-Verlag Berlin Heidelberg 2010

– Block sections for 20–30 min at room temperature before primary antibody incubation
Note: *Do not use sodium azide when working with HRP-conjugated antibodies.

When working with primary antibodies that come from mice and rabbits, one can also use 1% bovine serum albumin (BSA) solved in a buffer (PBS or TBS). We routinely use BSA-c basic blocking solution (1:10 in PBS; Aurion, Wageningen, The Netherlands). However, BSA as well as dry milk may contain bovine IgG. Many secondary antibodies such as anti-bovine, anti-goat, and anti-sheep will react strongly with bovine IgG. Therefore, use of BSA or dried milk for blocking or diluting these antibodies may significantly increase background and/or reduce antibody titer.

Never block with normal serum from the host species of the primary antibody. If immunoglobulins in the normal serum from the host species of the primary antibody bind to the specimen of interest, they will be recognized by the labeled secondary antibody, resulting in higher background.

In most cases, however, blocking endogenous Fc receptors is nothing more than a ritual. There are no persuasive studies demonstrating that the blocking serum step is really necessary. In any case, the use of labeled $F(ab')_2$ or $F(ab)$ fragments for the secondary antibodies in place of the whole IgG form of the secondary antibody eliminates the need to block endogenous Fc receptors (see Sect. 4.3.2).

5.2 Endogenous Peroxidase

Using HRP as an enzyme marker may result in high, non-specific background staining of some cells and tissues containing endogenous peroxidase. This non-specific background can be eliminated by pre-treatment of cells and tissues sections with hydrogen peroxide (H_2O_2).

Peroxidase Blocking Solution (0.6% H_2O_2 in Methanol or PBS)	
30% H_2O_2	1 ml
Methanol or PBS	50 ml
	Mix well and store at 4°C

Block sections for 10–15 min before or after primary antibody incubation.

Methanol is a better choice in most cases. Although methanol may reduce the staining of some cell surface antigens, such as CD4 and CD8, you do nothing wrong, even for these sensitive antigens, if you use peroxidase blocking solution with methanol after (not before) incubation with primary antibodies. Methanolic treatment may also cause frozen sections to detach from slide. In such rare cases, using hydrogen peroxide in PBS is recommended.

5.3   Endogenous Alkaline Phosphatase

Alkaline phosphatase is an enzyme represented by various isoforms in many tissues such as liver, bone, intestine, placenta, some tumors and in leukocytes. Addition of 1 mM levamisole to the chromogen/substrate will inhibit endogenous alkaline phosphatase activity, with the exception of the intestinal isoform. If necessary, this can be blocked with a weak acid wash, such as 0.03–0.5 N HCl or 1 M citric acid.

1 M citric acid

1. Dissolve 192 g citric acid (free acid) in 500 ml of distilled water.
2. Dilute to 1 liter with distilled water.

When using high-temperature antigen retrieval, you may skip blocking alkaline phosphatase, since some endogenous enzymes, such as alkaline and acid phosphatases, in contrast to peroxidase, are destroyed by boiling for even a short time at 100°C (Cattoretti et al. 1993).

5.4   Endogenous Biotin

Biotin, a B vitamin, present in some organs, such as kidney, liver and spleen, may be a cause high, non-specific background staining when using ABC method (see Sect. 6.2.1). To prevent this background staining, an avidin/biotin block can be applied to tissue sections containing moderate to high amounts of this vitamin prior to the incubation with biotinylated antibody. Blocking endogenous biotin can be done with commercially available kits, but they are expensive. You may, however, buy the individual components of the kits, free biotin and free avidin.

Avidin/Biotin Block

(A) Avidin 0.01% ($\approx$1.7 μM) in PBS
(B) Biotin 0.05% ($\approx$2 mM) in PBS
      Store these blocking solutions at 4°C.
      Incubate sections before primary antibody application with solutions A and B for 10–15 min each and rinse twice with PBS between steps.

Usually, avidin/biotin block is avoidable when working with paraffin tissue sections, since tissue pretreatment in this case normally leads to complete loss of endogenous biotin. You may, however, also stumble upon significant endogenous biotin artifacts in paraffin sections when subjecting your specimen to high-temperature antigen retrieval, and especially when employing signal amplification with biotinylated tyramide.

A cheaper practical alternative to this blocking step, with dilute egg whites as an avidin source, was suggested by Miller and Kubier (1997). This endogenous biotin-blocking procedure includes a distilled water rinse, incubating the slides in dilute egg white solution (two egg whites diluted to 200 ml with distilled water), followed by a rinse in distilled water, and finally an incubation in either 5% dried milk solution (25 g dried milk in 500 ml PBS/Tween buffer) or 0.2% biotin in PBS.

5.5 Autofluorescence or "The Wood Through the Trees"

Some specimens naturally fluoresce when illuminated by light of the proper wavelength. This phenomenon is called autofluorescence or primary fluorescence. In mammalian tissues, natural fluorescence is due in large part to substances such as flavins and porphyrins, lipofuscins (the breakdown product of old red blood cells — an "ageing" pigment; also prominent in certain large neurons in the CNS), elastin and collagen (connective tissue components, particularly in blood vessel walls) and some others. The causes of autofluorescence are different, but they have one thing in common: they complicate the use of fluorescence microscopy. Because of its broad excitation and emission spectra, autofluorescence overlaps with working spectra of most commonly used fluorophores. For a more detailed review of this

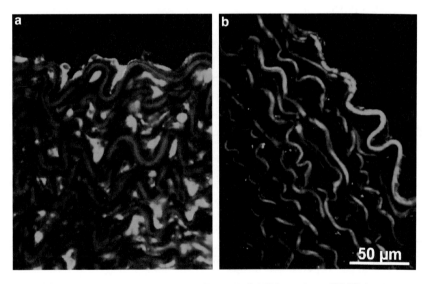

Fig. 5.1 Immunofluorescent demonstration of endothelial NO synthase (NOS3) in the porcine carotid artery. (**a**) FITC-tyramide (*green channel*) marks NOS3 in the media and intima. Lilac color of elastic laminae results as a sum of three images picked up under illumination with three filters exciting the fluorescence in the green, red and blue spectra. Separate images are captured digitally into color-separated components using an AxioCam digital microscope camera and AxioVision multi channel image processing (Carl Zeiss Vision GmbH, Germany). This approach allows discrimination between specific immunolabeling and non-specific autofluorescence emitted by elastic laminae shown in the control preparation (**b**) captured with the filter exiting the fluorescence in the green spectrum. Reproduced from Buchwalow et al. (2002)

subject, readers are referred to Billinton and Knight (2001) and to the webpage: http://www.uhnres.utoronto.ca/facilities/wcif/PDF/Autofluorescence.pdf.

Various solutions have been proposed for the reduction or elimination of auto-fluorescence. One way is to chemically suppress the autofluorescence signal with some reagents such as sodium borohydride, glycine or toluidine blue. However, in many cases, these approaches are either infeasible or ineffective, and none of them fully eliminates the problem. The second way is to use spectral unmixing algorithms subtracting the background fluorescence. This is only possible if you have at your disposal complicated, expensive confocal optics with sophisticated automated software (http://www.cri-inc.com/applications/fluorescence-microscopy.asp).

The pragmatic solution to this boring problem is to bring into play the fact that the autofluorescence spectrum is generally much broader than that of most commonly used fluorophores. Using a standard fluorescence microscope equipped with a digital camera, you can capture the autofluorescence in the spectrum region differing from that of your fluorescent label and merge them into a composite image. This will help you to convert the annoying autofluorescent background into a decoration for your immunolabeling results. For instance, elastic lamellae in the

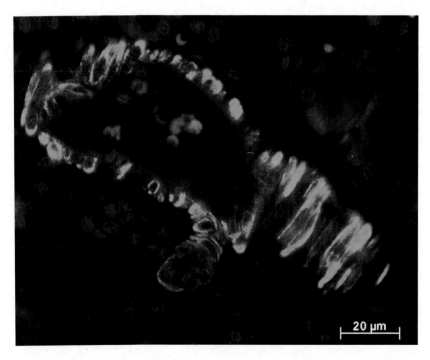

Fig. 5.2 Immunofluorescent demonstration of smooth muscle actin (FITC, *green channel*) in the blood vessel wall of the human kidney. Red autofluorescence of erythrocytes, elastic lamellae and kidney tubules was captured with a filter exciting the autofluorescence in red spectrum under a longer exposure than with the filter exciting specific fluorescence in the green spectrum. Nuclei are counterstained with DAPI (*blue channel*)

blood vessel wall elicit strong fluorescence in a broad spectrum from ~350 to 650 nm, covering the working region of most commonly used fluorophores. You can overcome this technical problem by triple exposure of the specimen to illumination with three filters exciting the fluorescence in the blue, green and red spectra (350 nm, 490 nm and 550 nm, respectively). The red, blue, and green components can be merged, and the elastic lamellae appear in a beautiful lilac color easily distinguishable from the specific immunolabeling (Fig. 5.1).

You can also capture the non-specific autofluorescence emitted by elastic lamellae and erythrocytes using a filter exciting the autofluorescence in the red spectrum under a longer exposure than with the filter exciting specific fluorescence in the green spectrum and merge them into a composite image, as shown in Fig. 5.2.

The simplest way, however, is to outsmart autofluorescence by using a wavelength for excitation that is out of the range of natural autofluorescence. The longer wavelength excitation of Cy5 or Cy7 in the infrared spectrum region can be chosen to avoid autofluorescence which is annoying only at shorter wavelengths.

References

Billinton N, Knight AW (2001) Seeing the wood through the trees: a review of techniques for distinguishing green fluorescent protein from endogenous autofluorescence. Anal Biochem 291:175–197

Buchwalow IB, Podzuweit T, Bocker W, Samoilova VE, Thomas S, Wellner M, Baba HA, Robenek H, Schnekenburger J, Lerch MM (2002) Vascular smooth muscle and nitric oxide synthase. FASEB J 16:500–508

Cattoretti G, Pileri S, Parravicini C, Becker MH, Poggi S, Bifulco C, Key G, D'Amato L, Sabattini E, Feudale E (1993) Antigen unmasking on formalin-fixed, paraffin-embedded tissue sections. J Pathol 171:83–98

Miller RT, Kubier P (1997) Blocking of endogenous avidin-binding activity in immunohistochemistry — the use of egg whites. Appl Immunohistochem 5:63–66

Chapter 6
Immunostaining Enhancement

In the last decade, pioneering efforts of histochemists have led to an immense improvement in reagents and protocols. Milestones in this development were antigen retrieval technique and signal amplification. Wide application of these techniques in pathology and other fields of morphology has demonstrated distinct enhancement of immunostaining on archival formalin-fixed, paraffin-embedded tissue sections for a variety of antigens. Whereas heat-induced antigen retrieval on formaldehyde-fixed and paraffin-imbedded tissues is used in the predetection phase, signal amplification with polymeric detection systems (e.g., EnVision System from Dako) and catalyzed reporter deposition (CARD) with tyramide are accomplished in detection and the post-detection phases respectively (Shi et al. 2001).

6.1 Antigen Retrieval

Antigens are affected differently by the various methods of pretreatment. Routine formaldehyde fixation and the rest of the procedure for paraffin embedding alter the conformation of protein macromolecules, negatively affecting antigen–antibody interaction and consequently decreasing the intensity of the final reaction in the immunohistochemical procedure. Whereas the conformation of the antibody used in the antigen–antibody reaction is usually preserved in its native form, the conformation of the antigen protein, located in the tissue, cannot be considered intact. It has suffered the effect of the aldehyde fixative, and may have been modified in its conformation or even in its constitution (Montero 2003). To recover the antigenicity of tissue sections that had been masked by formalin fixation and paraffin imbedding, tissue sections must be subjected to antigen retrieval (Hyatt 2002).

I.B. Buchwalow and W. Böcker, *Immunohistochemistry: Basics and Methods*,
DOI 10.1007/978-3-642-04609-4_6, © Springer-Verlag Berlin Heidelberg 2010

6.1.1 Heat-Induced Antigen Retrieval

Antigen–antibody recognition is dependent on protein structure. A conformational change in a protein caused by formalin fixation may mask the epitope and thus affect the antigenicity of proteins in formalin-treated tissue (Montero 2003). The antigen retrieval leads to a renaturation or at least partial restoration of the protein structure, with re-establishment of the three-dimensional protein structure to something approaching its native condition (Shi et al. 1991).

In the current literature, the term "antigen retrieval" is predominantly defined as a heat-induced antigen retrieval method. Development of the heat-induced antigen retrieval technique, a simple method of boiling archival paraffin-embedded tissue sections in buffered water solutions to enhance the signal of immunohistochemistry, rendered immunohistochemistry applicable to routine formalin-fixed, paraffin-embedded tissues for wide application in research and clinical pathology (Shi et al. 2001). This technique has demonstrated distinct enhancement of immunostaining on archival formalin-fixed, paraffin-embedded tissue sections for a variety of antigens, such as Ki-67, MIB1, estrogen and androgen receptors, many cytokeratin and CD markers, which often appear otherwise negative in immunohistochemistry.

For antigen retrieval, tissue sections are placed in a container with an antigen retrieval solution and heated at 90°C–110°C (depending on tissue). Initially described using microwave as a source of heat, this technique has been also found to be effective when employing other heat sources, including autoclaves (Bankfalvi et al. 1994), domestic pressure cookers (Buchwalow et al. 2002), hot water baths and vegetable steamers. The choice of an antigen retrieval solution, the temperature and length of pretreatment will vary for different antigens and different tissues, as well as for the fixation processes that have been utilized. In our laboratory, we routinely use domestic vegetable steamers with the temperature of retrieval solution around 95°C and incubation time usually 30 min.

Antigen retrieval using a domestic vegetable steamer or a water bath
Preheat steamer or water bath with a Coplin jar filled with an antigen retrieval solution of choice until temperature reaches 95°C–100°C. Immerse slides in the Coplin jar and place the lid loosely on the jar. Incubate for 30–60 min and turn off the steamer or water bath. Place the Coplin jar at room temperature and allow the slides to cool for 20 min.

Antigen retrieval using the microwave method
Before the beginning of the immunostaining protocol, slides should be put in a Coplin jar filled with an antigen retrieval solution of choice and heated in a commercial microwave oven operating at a frequency of 2.45 GHz and 600 W power setting. After two heating cycles of 5 min each, slides should be allowed to cool at room temperature and thoroughly washed in PBS.

Ready-to-use antigen retrieval solutions are commercially available from Biocarta (Hamburg, Germany), Dianova (Hamburg, Germany), Serotec (Dusseldorf, Germany) and other manufacturers. Available from Dako (Dako, Glostrup, DK),

Target Retrieval Solutions from DAKO with pH 6.0 or pH 9.0 are well-suited for use on formalin-fixed, paraffin-embedded tissue sections mounted on glass slides. Compared with 0.01 M Citrate buffer, pH 6, the use of 0.001 M EDTA buffer, pH 9, significantly improves staining results for many antigens, preserves the morphology better and is especially useful in combination with the Dako EnVision visualization systems.

Alternatively, customers may prefer to prepare their own buffers using the recipes provided below. There is a fairly wide variety of antigen retrieval solutions that may be employed, including citrate buffer (0.01–0.1 M, pH 6), EDTA (0.001–0.1 M, pH 8.0 or 9.0), Tris base buffer (0.5 M, pH 10), 0.05 M glycine-HCl buffer, 1% periodic acid, various concentrations of urea, lead thiocyanate solutions, and even distilled water or liquid hand soap. The choice of a preprepared antigen retrieval solution should be as recommended on product specific datasheets for antibodies, or determined experimentally by the customer. When other antigen retrieval methods fail, antigen retrieval can be accomplished in 90% glycerol solution using a hot plate with a magnetic stir rod. Glycerol has an advantage of having a very high boiling point (290°) and being nontoxic and reusable. The stir bar maintains a constant and uniform temperature throughout the antigen retrieval fluid. For an overview of antigen retrieval methods the reader is referred to Hyatt (2002).

Tris buffered saline (TBS) (stock solution × 10 concentrated)

Sodium chloride	87.66 g
Tris	60.55 g
Distilled water	1 l
Adjust pH to 7.4 using concentrated HCl	

0.01 M Citrate buffer for antigen retrieval

Anhydrous citric acid	3.84 g
Distilled water	1,800 ml
Adjust pH to 6.0 using concentrated NaOH.	
Make up to 2 l with distilled water	

0.001 M EDTA buffer for antigen retrieval

EDTA	0.37 g
Distilled water	1,000 ml
Adjust pH to 9.0 with 1 M NaOH	

Heat-induced antigen retrieval is not needed for immunostaining of fresh frozen sections fixed with acetone, since heat-induced antigen retrieval is used to break the cross-links formed by formalin fixation. Acetone-fixed frozen sections are fragile and are easily destroyed by heating. For aldehyde-fixed frozen material, antigen retrieval can be, however, achieved by heating tissue probes en bloc (Ino 2003). This simple method requires no special equipment, and is useful for researchers who prefer frozen sections or do not dispose of facilities for paraffin embedding.

Antigen retrieval by heating en bloc for prefixed frozen material (Ino 2003)

1. Prepare tissues fixed with 4% formaldehyde in 0.1 M sodium phosphate buffer, pH 7.5. The tissue blocks should be cut to a proper size (e.g., slices 3–5 mm thick).
2. Immerse the tissue blocks in a retrieval solution (distilled water or 10 mM sodium citrate buffer, pH 6.0) at 4°C overnight.
3. Place the tissue blocks in a small, heat-resistant basket and immerse in boiling retrieval solution (200–500 ml) with gentle stirring, using a hot plate for 3–5 min. For heating, a conventional burner can also be used.
4. Immediately place the tissue blocks in cold 30% sucrose in PBS, and incubate at 4°C until the blocks sink.
5. Immerse the tissue blocks in an embedding medium, and freeze quickly with crushed dry ice. The frozen tissue blocks can now be stored at −80°C.
6. Cut frozen sections with a cryostat and mount them on glass slides coated with freshly prepared 0.02% poly-L-lysine. Free-floating sections can also be used.
7. Dry the sections well for more than 1 day in a vacuum desiccator at RT. If no vacuum desiccator is available, dry for several days at RT to prevent detachment of the sections during processing. The dried sections can be stored in a desiccated chamber at −20°C for several months.

Another simple alternative method for antigen retrieval in aldehyde-fixed cryostat tissue sections and cell cultures involves treatment with sodium dodecyl sulfate (SDS) (Brown et al. 1996)

Antigen retrieval in aldehyde-fixed cryostat tissue sections and cell cultures by treatment with SDS (Brown et al. 1996)

1. Rinse sections three times for 5 min each in PBS.
2. Cover sections with 1% SDS in 0.01 M PBS (pH 7.4) solution and incubate for 5 min at room temperature.
3. Rinse sections three times for 5 min each in PBS. It is important to wash sections well, otherwise residual SDS will denature the antibodies subsequently applied to sections.
4. Incubate sections in serum blocking solution.
5. Incubate in the primary antibody and complete immunohistochemical staining steps as desired.

6.1.2 Proteolytic Antigen Retrieval

Antigenic determinants masked by formalin-fixation and paraffin-embedding may also be exposed by enzymatic digestion. This can, however, not be used with frozen sections or cells which are not paraffin-embedded. The beneficial effects of protease treatment are presumably related to cleavage of the molecular cross-links by the proteolytic enzyme, allowing the antigen to return to its normal conformation, which serves for more effective antibody binding. A wide variety of proteases have been employed, including trypsin, proteinase K, pepsin, pronase, ficin, and others. Concentration of enzyme is usually 0.05–0.1%, depending on type of tissue and fixation. Incubation time within 5–30 min is commonly used. Incubation temperature is usually at 37°C.

Proteolytic antigen retrieval using Trypsin
1. Dissolve 0.1 g calcium chloride in 100 ml distilled water and adjust pH to 7.8 with 0.1 M sodium hydroxide. Store at 37°C.
2. Dissolve 0.1 g trypsin (Sigma Type II) in calcium chloride solution.
3. Place sections in trypsin solution at 37°C and incubate for predetermined optimum time (approximately 20–30 min).
4. Wash in TBS and proceed with staining.

Proteolytic antigen retrieval using pronase
1. Dissolve 0.1 g calcium chloride in 100 ml distilled water.
2. Dissolve 0.1 g pronase in calcium chloride solution, and adjust pH to 7.8 with 0.1 M sodium hydroxide.
3. Place sections in pronase solution at room temperature, and incubate for predetermined optimum time (approximately 10 min).
4. Wash in TBS and proceed with staining

Unfortunately, enzymatic digestion fails to yield satisfactory immunostaining for many antigens, not to say that enzymatic digestion of tissue sections may negatively affect the structural integrity of the specimen. With the advent of heat-induced antigen retrieval techniques, earlier unmasking techniques with protease digestion play a much smaller role. For more information about antigen retrieval solutions the reader is referred to the website: http://www.ihcworld.com/.

6.2 Signal Amplification

Whereas antigen-retrieval technique serves to amplifying the immunocytochemical signal at the predetection phase, conventional methods of signal amplification, such as avidin–biotin complex (ABC) and soluble enzyme-anti-enzyme immune complex techniques (peroxidase-anti-peroxidase complex and alkaline phosphatase-anti-alkaline phosphatase complex — PAP and APAAP respectively), are applied in the phase of detection. For many years, the PAP and APAAP procedures represented the most sensitive and reliable and hence most popular techniques in many pathology laboratories. However, today these techniques are only rarely used, being substituted by modern more sensitive methods.

6.2.1 Avidin–Biotin Complex

Over the years, several ABC techniques have evolved. The ABC method is based on the high affinity that streptavidin (from *Streptomyces avidinii*) and avidin (from chicken egg) have for biotin. Biotin is a naturally occurring vitamin (vitamin B7, vitamin H). One mole avidin will bind four moles biotin. The avidin–biotin interaction is the strongest known noncovalent, biological interaction. The bond formation is rapid and is unaffected over wide range of pH. Because avidin has a propensity to nonspecifically bind to lectin-like and negatively charged tissue components at physiological pH, it has been largely replaced today by streptavidin. By covalently linking (strept)avidin with different haptens, such as fluorophores, enzymes or colloidal gold, the ABC system can be utilized in light, fluorescence and electron microscopy. The basic sequence of ABC reagents application consists of (1) primary antibody directed toward a specific determinant on the cells, (2) biotinylated secondary antibody, and (3) labeled (strept)avidin. The most recent and the most widely used ABC technique is a patented procedure called the "preformed complex" method (http://www.vectorlabs.com/). After application of a biotinylated secondary or primary antibody, a preformed complex between streptavidin and a biotinylated enzyme is added. This latest technique appears to be the most sensitive in many ABC applications. The enhanced sensitivity is particularly important in the localization of antigens present in low amounts or in cases where the cost of primary antibodies is significant. The increased sensitivity also provides an option to substantially reduce staining times.

Staining procedure for paraffin sections using ABC system
1. Deparaffinize and rehydrate tissue sections. Rinse in distilled water for 5 min.
2. Perform antigen retrieval if desired (see Sect. 6.1.1).
3. When using HRP as an enzyme marker, incubate the sections for 15 min in 0.3% H_2O_2 in either methanol or water to quench endogenous peroxidase (see Sect. 5.2). If endogenous peroxidase activity does not present a problem, this step may be omitted.
4. Wash in buffer for 5 min (see Sect. 3.4).
5. Incubate sections for 20 min with PBS containing 5% normal serum of species in which the secondary antibodies were raised.
6. Blot excess serum from sections.
7. Endogenous biotin may be a cause for nonspecific background staining (see Sect. 5.4). To eliminate this unwanted staining, apply an avidin/biotin block (for instance Avidin/Biotin blocking kit from VECTASTAIN, Cat. No. SP-2001). Usually, paraffin tissue sections are free from endogenous biotin, and this step may be omitted.
8. Incubate sections for 30–60 min at room temperature or overnight at +4°C with primary antibody diluted in buffer. If your primary antibody is biotinylated, you may omit the next two steps.

9. Wash slides for 3 × 3 min in buffer.
10. Incubate sections for 30 min with correspondingly diluted biotinylated secondary antibody.
11. Wash slides for 3 × 3 min in buffer.
12. Incubate sections for 30 min with (strept)avidin conjugated with a fluorophore or with any enzyme (peroxidase or alkaline phosphatase). For localization of low-density antigens (<10K molecules/cell), VECTASTAIN ABC reagent (http://www.vectorlabs.com/) is preferable.
13. Wash slides for 3 × 3 min in buffer.
14. When using an enzyme (peroxidase or alkaline phosphatase) marker, incubate sections in an enzyme substrate solution until desired stain intensity develops (see Sect. 2.3). A fluorophore label can be directly visualized in a fluorescent microscope.
15. Counterstain nuclei if necessary, e.g., with DAPI for fluorescence microscopy or with hematoxylin for brightfield microscopy.
16. When using a fluorescent label, a short treatment (1–3 min) with 4% formaldehyde in PBS before mounting in water-soluble media is recommended to block the detachment of the fluorophore from the antibody, this preserves the staining patter for a longer storage.
17. Wash sections for 3 × 3 min in buffer.
18. Mount sections in aqueous medium or balsam for brightfield microscopy or in anti-fade medium for fluorescence microscopy (see Sect. 3.5).

All incubations are at room temperature unless otherwise noted.

Convenient ready-to-use ABC systems can also be purchased from DAKO (http://www.dako.com/), DCS Innovative Diagnostik Systeme (http://www.dcs-diagnostics.com/) and from some other vendors. The website: http://www.dcs-diagnostics.de/data/IHC_Arbeitsanleitung_AP_web.pdf describes the staining procedure for paraffin sections using ABC system in the German language.

6.2.2 Chain Polymer-Conjugated Technology

A several-fold higher antigen detectability than those achieved in the standard ABC protocols or enzyme-anti-enzyme immune complex techniques can be gained with the chain polymer-conjugated technology (EnVision System) developed by Dako-Cytomation. This technology utilizes an enzyme-labeled inert "spine" molecule of dextran. In addition to an average of 70 molecules of enzyme (AP or HRP), ten molecules of antibody can be attached to the spine molecule (see Fig. 6.1). Because this system avoids the use of (strept)avidin and biotin, nonspecific staining as a result of endogenous biotin is eliminated.

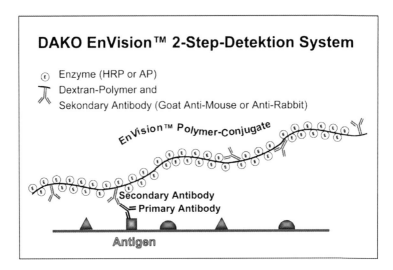

Fig. 6.1 Chain polymer-conjugated technology (EnVision System) developed by DakoCytomation

6.2.3 Tyramide Signal Amplification

With the recent introduction of tyramine conjugates (tyramides) as substrates for horseradish peroxidase (HRP), the sensitivity of immunohistochemical methods is immensely improved, and possible applications of IHC are extended to new limits (Adams 1992; Bobrow et al. 1989; Buchwalow 2002; Hopman et al. 1998; van Gijlswijk et al. 1997). This procedure, designated as a CARD or tyramide signal amplification (TSA), takes advantage of HRP from an HRP-labeled secondary antibody or from an HRP–avidin complex to catalyze in the presence of hydrogen peroxide the oxidation of the phenol moiety of labeled tyramine. On oxidation by HRP, activated tyramine molecules rapidly bind covalently to electron-rich amino acids of proteins immediately surrounding the site of the immunoreaction (see Fig. 6.2). This allows an increase in the detection of an antigenic site up to 100-fold compared with the conventional indirect method, with no loss in resolution. Wide application of these pioneering techniques in pathology and other fields of morphology has demonstrated distinct enhancement of immunohistochemical staining on archival formalin-fixed, paraffin-embedded tissue sections for a variety of antibodies due to reduction of the detection thresholds (increasing sensitivity).

The TSA technology was developed by Litt's group (Bobrow et al. 1989) at DuPont NEN (now a part of PerkinElmer Corporation) and licensed to Molecular Probes for in-cell and in-tissue applications. For approximately € 200.00, a single TSA kit can be purchased that is enough for approximately 75–100 slides. However, synthesis of these reagents in-house is not only simple but can potentially save a lot of money. For less than € 50.00, reagents can be purchased to stain over 100,000 slides using a procedure for the synthesis of different tyramide conjugates

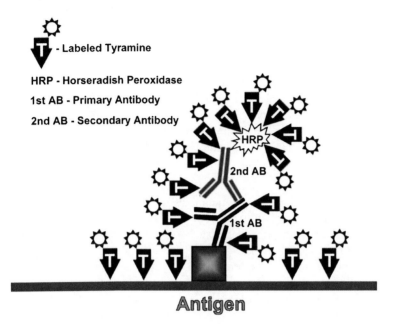

Fig. 6.2 Tyramide signal amplification. "T" is the labeled tyramine and "HRP," horse radish peroxidase. The "Label" can be a fluorochrome or biotin. The fluorochrome can be visualized directly in a fluorescence microscope. Biotin can be visualized via labeled streptavidin

reported by Hopman et al. (1998). According to this one-step procedure, succinimidyl esters of biotin, digoxigenin, or fluorophores are coupled to tyramine in dimethylformamide (DMF) adjusted to a pH of 7.0–8.0 with triethylamine (TEA). The coupling reaction can be performed within 2 h, and the reaction mixture can be applied without further purification steps. The synthesized tyramide conjugates can also be applied in in situ hybridization procedures to detect both repetitive and single-copy DNA target sequences in cell preparations with high efficiency. This approach provides an easy and fast method to prepare a variety of tyramide conjugates in bulk amounts for yourself and your friends at rather low cost. But you can not sell it. It is patented.

FITC-Tyramide synthesis
1. Active ester stock solution "A": dissolve10 mg of 5-(and-6)carboxyfluorescein, succinimidyl ester (FITC-NHS, Molecular Probes #C-1311) in 1 ml of DMF.
2. Tyramine-HCl stock solution "B": dissolving 10 mg of Tyramine-HCl (Sigma, T-2879) in 1 ml of DMF to which 10 µl of triethylamine (TEA, Pierce) was added.
3. For efficient acylation, mix 340 µl of stock solution B with 1 ml of stock solution A, and leave at room temperature in the dark for 2 h.
4. The synthesized tyramide conjugates are diluted in 10 ml 100% methanol. The tyramide stock solutions may be safely stored for at least 8 months at 4°C without any reduction in reactivity.

Biotin-Tyramide synthesis
1. Active ester stock solution "A": dissolve 10 mg of sulfosuccinimidyl-6-(biotini-mide)hexanoate (BIO-NHS, Pierce) in 1 ml of DMF.
2. Tyramine-HCl stock solution "B": dissolve 10 mg of Tyramine-HCl (Sigma, T-2879) in 1 ml of DMF to which 10 µl of triethylamine (TEA, Pierce) has been added.
3. For efficient acylation, mix 289 µl of stock solution B with 1 ml of stock solution A and leave at room temperature in the dark for 2 h.
4. The synthesized tyramide conjugates are diluted in 10 ml 100% ethanol. The tyramide stock solutions may be safely stored for at least 8 months at 4°C without any reduction in reactivity.

Alternatively, biotinylation of tyramine in aqueous solutions can be carried out according to the protocol of Adams (1992). In brief, tyramine hydrochloride (6.2 mg; Sigma, Deisenhofen, Germany, T-2879) is diluted in 50 mM borate buffer (8 ml, pH 8.0), containing sulfo-NHS-LC-Biotin (20 mg; Pierce, Rockford, IL, USA), and biotinylation is performed overnight under continuous stirring. The solution is filtered (0.45 syringe filter), and 5 µl aliquots are stored ($-70°C$) until use (von Wasielewski et al. 1997). The tyramide solution is applied in a dilution 1:2,000 in 0.3% H_2O_2/TBS for 5 min, then allowing a fivefold increase in antibody dilution.

Immunofluorescence staining using FITC-Tyramide
1. Deparaffinize and rehydrate tissue sections. Rinse in distilled water for 5 min.
2. Perform antigen retrieval if desired (see Sect. 6.1.1).
3. Washes and dilutions: Wash sections in 10 mM sodium phosphate buffer, pH 7.5, 150 mM NaCl (PBS) for 2×3 min. PBS is used for all washes and dilutions. Other buffers such as TBS may also be used.
4. Blocking step 1: to block endogenous HRP activity incubate sections in 0.3% H_2O_2 in methanol for 15 min (see Sect. 5.2) and wash in PBS for 3×3 min.
5. Blocking step 2: to block endogenous Fc receptor, incubate sections for 20 min with PBS containing 5% normal serum of species in which the secondary antibodies were raised.
6. Primary antibodies: blot excess blocking solution from sections and incubate for 60 min at room temperature or over night at $+4°C$ with a correspondingly diluted primary antibody. Wash sections in PBS for 3×3 min.
7. Secondary antibodies: incubate sections for 60 min at room temperature with a secondary HRP-conjugated antibody raised against the corresponding IgG of the primary antibody. Using DAKO HRP-EnVision System instead of HRP-conjugated antibody at this step will contribute for further powerful signal amplification. Wash sections in PBS for 3×3 min.
8. Signal amplification: incubate sections for 8–10 min at room temperature with FITC-Tyramide dissolved 1:500 in amplification diluent (0.02% H_2O_2 in PBS)

to achieve the final concentration of the active tyramide 11.6 µM and wash in PBS for 3 × 3 min.

9. Counterstaining: Counterstain nuclei if necessary, e.g., with DAPI (Sigma, Cat. 18860, 5 µg/ml PBS). Wash sections in PBS for 3 × 3 min.

10. Mounting: mount sections in an anti-fade medium for fluorescence microscopy. The authors have a good experience with Vectashield mounting medium from Vector Laboratories (http://www.vectorlabs.com/).

All incubations are at room temperature unless otherwise noted.

Immunofluorescence staining using biotinylated Tyramide

1. Deparaffinize and rehydrate tissue sections. Rinse in distilled water for 5 min.
2. Perform antigen retrieval if desired (see Sect. 6.1.1).
3. Washes and dilutions: wash sections in 10 mM sodium phosphate buffer, pH 7.5, 150 mM NaCl (PBS) for 2 × 3 min. PBS is used for all washes and dilutions. Other buffers such as TBS may also be used.
4. Blocking step 1: to block endogenous HRP activity incubate sections in 0.3% H_2O_2 in methanol for 15 min (see Sect. 5.2) and wash in PBS for 3 × 3 min.
5. Blocking step 2: to block endogenous Fc receptor, incubate sections for 20 min with PBS containing 5% normal serum of species in which the secondary antibodies were raised.
6. Primary antibodies: blot excess blocking solution from sections and incubate for 60 min at room temperature or over night at +4°C with a correspondingly diluted primary antibody. Wash sections in PBS for 2 × 3 min.
7. Secondary antibodies: incubate sections for 60 min at room temperature with a secondary HRP-conjugated antibody raised against the corresponding IgG of the primary antibody. Using DAKO HRP-EnVision System instead of HRP-conjugated antibody at this step will contribute for a further powerful signal amplification. Wash sections in PBS for 3 × 3 min.
8. Signal amplification: incubate sections for 8–10 min at room temperature with biotinylated Tyramide in amplification diluent (0.02% H_2O_2 in PBS) dissolved 1:1,500 to achieve the final concentration of the active tyramide 3.87 µM and wash in PBS for 2 × 3 min.
9. Visualization of biotin: biotin can be visualized using (strept)avidin conjugated with a fluorophore of with an enzyme label.
10. Counterstaining: Counterstain nuclei if necessary, e.g., with hematoxylin for brightfield microscopy or with DAPI (Sigma, Cat. 18860, 5 µg/ml PBS) for fluorescence microscopy. Wash sections in PBS for 2 × 3 min.
11. Mounting: mount sections in water-soluble medium or balsam for brightfield microscopy or in an anti-fade medium for fluorescence microscopy.
12. All incubations are at room temperature unless otherwise noted.
13. Note: because of a dramatic increase in the sensitivity this method may require an additional blocking step for inactivation of endogenous biotin in some tissue specimens (see Sect. 5.4).

6.2.4 Amplification of the Amplifier

Whereas the fluorochrome label can be visualized directly in a fluorescence microscope, the biotin label can be further amplified via ABC technique with HRP-conjugated avidin or streptavidin, and visualized via chromogenic enzymo-cytochemical reaction using brightfield microscopy. This second-layer HRP label can be further amplified with the second-layer fluorochrome–tyramine. Alternatively, ABC amplification can be introduced after the second step of the immuno-cytochemical reaction, employing secondary antibodies conjugated with biotin instead of HRP. These combinations of tyramide-based amplification with ABC method are called a "blast" amplification technique, which allows a further signal enhancement up to 1,000-fold. TSA can be also successfully combined with the chain polymer-conjugated technology (HRP-EnVision System, DAKO). It makes it possible to detect antigens with a very low level of expression, where the conventional indirect methods failed. Moreover, the amplification requires that the primary antibody be greatly diluted. This dilution results in less background staining, and yet stronger signals are produced.

References

Adams JC (1992) Biotin amplification of biotin and horseradish-peroxidase signals in histochemical stains. J Histochem Cytochem 40:1457–1463

Bankfalvi A, Navabi H, Bier B, Bocker W, Jasani B, Schmid KW (1994) Wet autoclave pretreatment for antigen retrieval in diagnostic immunohistochemistry. J Pathol 174:223–228

Bobrow MN, Harris TD, Shaughnessy KJ, Litt GJ (1989) Catalyzed reporter deposition, a novel method of signal amplification — application to immunoassays. J Immunol Meth 125:279–285

Brown D, Lydon J, McLaughin M, StuartTilley A, Tyszkowski R, Alper S (1996) Antigen retrieval in cryostat tissue sections and cultured cells by treatment with sodium dodecyl sulfate (SDS). Histochem Cell Biol 105:261–267

Buchwalow IB, Podzuweit T, Bocker W, Samoilova VE, Thomas S, Wellner M, Baba HA, Robenek H, Schnekenburger J, Lerch MM (2002) Vascular smooth muscle and nitric oxide synthase. FASEB J 16:500–508

Hyatt MA (2002) Microscopy, Immunohistochemistry, and Antigen Retrieval Methods: For Light and Electron Microscopy. Plenum, New York

Hopman AHN, Ramaekers FCS, Speel EJM (1998) Rapid synthesis of biotin-, digoxigenin-, trinitrophenyl-, and fluorochrome-labeled tyramides and their application for in situ hybridization using CARD amplification. J Histochem Cytochem 46:771–777

Ino H (2003) Antigen retrieval by heating en bloc for pre-fixed frozen material. J Histochem Cytochem 51:995–1003

Montero C (2003) The antigen–antibody reaction in immunohistochemistry. J Histochem Cytochem 51:1–4

Shi SR, Key ME, Kalra KL (1991) Antigen retrieval in formalin-fixed, paraffin-embedded tissues: an enhancement method for immunohistochemical staining based on microwave oven heating of tissue sections. J Histochem Cytochem 39:741–748

Shi SR, Cote RJ, Taylor CR (2001) Antigen retrieval techniques: current perspectives. J Histochem Cytochem 49:931–938 van Gijlswijk RPM, Zijlmans HJMA, Wiegant J, Bobrow MN, Erickson TJ, Adler KE, Tanke HJ, Raap AK (1997) Fluorochrome-labeled tyramides: use in immunocytochemistry and fluorescence in situ hybridization. J Histochem Cytochem 45:375–382

von Wasielewski R, Mengel M, Gignac S, Wilkens L, Werner M, Georgii A (1997) Tyramine amplification technique in routine immunohistochemistry. J Histochem Cytochem 45: 1455–1460

Chapter 7
Multiple Multicolor Immunoenzyme Staining

Immunoenzyme staining can be utilized for simultaneous localization of two or more tissue antigens in cases when the antigens of interest are located in different cellular compartments or in different cells. This approach is, however, inapplicable for co-localization of a pair of antigens located in the same cellular compartment, because of massive deposits of chromogen molecules covering the antigen targeted by the enzyme marker. This renders the second antigen of interest inaccessible for the second primary antibody, and makes the targeting of a pair of antigens in the same cellular compartment (nucleus, membrane, cytoplasm) impossible. In such situations, one should use immunofluorescent labeling with fluorophore-conjugated antibodies (see Chap. 8).

A collection of useful protocols for the simultaneous immunoenzyme staining of two or more tissue antigens is available on the websites: http://www.protocol-online.org/prot/Immunology/; http://www.vectorlabs.com/Protocols/MLB.pdf and http://www.ihcworld.com/. The protocols given in this chapter are practiced in the authors' laboratory. These methods fall into two main categories: (a) simultaneous immunoenzymatic double staining, and (b) sequential immunoenzymatic double/multiple staining.

7.1 Simultaneous Immunoenzymatic Double Staining

Simultaneous immunoenzymatic double staining is applicable for co-localization of two different antigens with a pair of primary antibodies raised in two different species. This approach can also be utilized for a pair of primary antibodies raised in the same species, provided that they are of different IgG isotypes (e.g., IgG1 and IgG3). A pair of primary antibodies of different IgG species or different IgG isotypes of the same species is applied to a tissue section as a mixture with the following visualization, using a mixture of a pair of secondary enzyme-labeled antibodies against corresponding IgG species or IgG isotypes.

I.B. Buchwalow and W. Böcker, *Immunohistochemistry: Basics and Methods*,
DOI 10.1007/978-3-642-04609-4_7, © Springer-Verlag Berlin Heidelberg 2010

Simultaneous immunoenzymatic double staining requires the use of two different enzyme systems. The most frequently used enzyme systems are HRP and AP (see Sect. 2.3). AP substrates form precipitates that are generally not as dense as that achieved with a HRP reaction. Therefore HRP labeling, as a rule, allows a better resolution of an antigenic site than AP. In double-label immunostaining, the more stable HRP label should be developed before AP. The most abundant antigen should be stained first, followed by staining of the least abundant antigen.

Basic protocol for simultaneous immunoenzymatic double staining using primary antibodies of two different species or two different IgG isotypes

1. Deparaffinize and rehydrate tissue sections. Rinse in distilled water for 5 min.
2. Antigen retrieval: place sections in a Coplin jar with antigen retrieval solution of choice (e.g., 10 mM citrate acid, pH 6) and heat at 90–110°C (depending on tissue) in a microwave, steamer, domestic pressure cooker or autoclave (see Sect. 6.1.1).
3. Washes and dilutions: wash sections in PBS or TBS for 2×3 min. PBS can be used for all washes and dilutions. When working with AP-enzyme system, TBS must be used.
4. Blocking step 1: to block endogenous enzyme activity or endogenous biotin (if required), use the corresponding blocking system (see Chap. 5) and wash in PBS or TBS for 2×3 min.
5. Blocking step 2: to block endogenous Fc receptor, incubate sections for 20 min with PBS or TBS containing 5% normal serum of species in which the secondary antibodies were raised.
6. Primary antibodies: blot excess blocking solution from sections and incubate for 60 min at room temperature or overnight at $+4$°C with a mixture of correspondingly diluted primary antibodies raised in two different species (e.g., rabbit and mouse) or belonging to two different IgG isotypes (e.g., IgG1 and IgG3). Wash sections in PBS or TBS for 3×3 min.
7. Secondary antibodies: incubate sections for 60 min at room temperature with a mixture of secondary antibodies conjugated with different enzymes (e.g., HRP and AP) and raised against the corresponding species IgG or IgG isotypes. Wash sections in a buffer recommended for developing the first enzyme label. The more stable HRP label should be developed before AP.
8. Substrate: incubate sections with appropriate enzyme substrate (see Sect. 2.3) until optimal color develops, monitor development with microscope. Wash sections first in the same buffer, and thereafter twice in a buffer recommended for developing the second enzyme (AP) label. Proceed with protocol to develop the second enzyme label.
9. Counterstaining and mounting: counterstain nuclei (see Sect. 7.4) and coverslip using the appropriate protocol for aqueous or permanent mounting, depending on the chemical nature of the developed chromogen (see Sect. 3.5 and consult the insert leaflet of the manufacturer).

Figure 7.1 illustrates the application of this technique with double immunolabeling of two antigens, Ki67 (Mouse IgG) and Von Willebrand Factor (F8, Rabbit IgG)

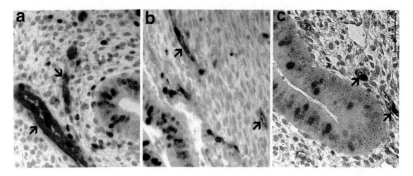

Fig. 7.1 Simultaneous localization of two antigens, Ki67 (Mouse IgG), as a proliferation marker, and Von Willebrand Factor (F8, Rabbit IgG), as an endothelial marker, in human endometrium. **a** Von Willebrand Factor is visualized with VECTOR Red substrate kit via goat-anti-rabbit AP-conjugated antibody and Ki67, with DAB substrate kit via goat-anti-mouse HRP-conjugated antibody. **b** Von Willebrand Factor is visualized with VECTOR Red substrate kit via goat-anti-rabbit AP-conjugated antibody and Ki67, with VECTOR SG substrate kit via goat-anti-mouse HRP-conjugated antibody. **c** Von Willebrand Factor is visualized with VECTOR SG substrate kit via goat-anti-rabbit HRP-conjugated antibody and Ki67, with VECTOR Red substrate kit via goat-anti-mouse AP-conjugated antibody

in human endometrium. Ki67 is a nuclear antigen present in proliferating human cells. Von Willebrand Factor antibody specifically reacts with the cytoplasm of human endothelial cells. For simultaneous immunostaining of these two antigens, the two primary antibodies (to Ki67 and to F8) and consequently the two corresponding secondary antibodies (to mouse IgG and to rabbit IgG) were applied as a mixture.

A ready-to-use system for double HRP and AP immunostaining using a pair of primary antibodies, which come from mouse and rabbit, can be purchased from DCS Innovative Diagnostik Systeme (http://www.dcs-diagnostics.com/).

7.2 Sequential Immunoenzymatic Double/Multiple Staining

When two primary antibodies are raised in the same species (a pair of mouse antibodies or a pair of rabbit antibodies) and are of the same IgG isotype, one uses the sequential immunoenzymatic double staining. This approach allows the sequential use also of the same enzyme systems with, however, different enzyme substrates (chromogens) in contrasting colors to label each antigen. This is possible because the first enzyme label gets masked through massive deposits of chromogen molecules covering the antigen targeted by this enzyme marker.

Basic protocol for sequential immunoenzymatic double staining (Adapted from http://www.vectorlabs.com/ and http://www.ihcworld.com/_books/Dako_Handbook.pdf)

1. Deparaffinize and rehydrate tissue sections. Rinse in distilled water for 5 min.
2. Perform antigen retrieval if desired (see Sect. 6.1.1).
3. Washes and dilutions: wash sections in PBS or TBS for 2 × 3 min. PBS can be used for all washes and dilutions. When working with AP-labeled secondary antibodies, TBS must be used.
4. Blocking step 1: to block endogenous enzyme activity or endogenous biotin, use the corresponding blocking system (see Chap. 5) and wash in PBS or TBS for 2 × 3 min.
5. Blocking step 2: to block endogenous Fc receptor, incubate sections for 20 min with PBS or TBS containing 5% normal serum of species in which the secondary antibodies were raised.
6. The first immunostaining sequence: blot excess blocking solution from sections, and incubate for 60 min at room temperature or overnight at +4°C with a correspondingly diluted first primary antibody. Wash sections in PBS or TBS for 3 × 3 min.
7. Secondary antibodies: incubate sections for 60 min at room temperature with the corresponding enzyme-conjugated secondary antibody. Wash sections in a buffer recommended for the corresponding enzyme substrate development.
8. Substrate: incubate sections with appropriate first enzyme substrate until optimal color develops (see Sect. 2.3). Wash sections first in the same buffer.
9. If the staining protocol must be interrupted, sections may be kept in a buffer solution at 4°C until staining is resumed.
10. The second immunostaining sequence: to stain the second antigen, incubate sections with the second primary antibody and thereafter with the second enzyme-conjugated secondary antibody. Visualize the second enzyme label with a chromogen developing a color different from the color of the first chromogen. Wash sections in a buffer recommended for developing the second enzyme label.
11. Counterstaining and mounting: counterstain nuclei (see Sect. 7.4) and coverslip using the appropriate protocol for aqueous or permanent mounting (see Sect. 3.5 and consult the insert leaflet of the manufacturer).

A third antigen can be detected (triple labeling) by simply continuing the same protocol with an additional staining sequence appropriate for the third primary antibody. Multiple labeling can increase the time required for staining antigens. If necessary, choose an appropriate stopping point during the staining protocol. If the staining protocol must be interrupted, it should be after substrate development of the first antigen. Slides may be kept in a buffer bath for up to 1 h at room temperature (20–25°C) without affecting the staining performance. For a longer time, sections can be kept in PBS at 4°C until staining is resumed (http://www. vectorlabs.com).

7.3 "Stripping" Buffers for Sequential Immunoenzymatic Double Staining

The success of the sequential immunoenzymatic procedure for double/multiple staining is based on the complete development of the first stain, which should mask the structure stained first. However, a danger of potential cross-reactivity still persists. An intermediary elution step using "Double Stain Block" developed by DAKO avoids this obstacle. Applied prior to the use of next primary antibodies, this step serves to remove previously bound primary and link antibodies, leaving only the deposit of chromogen from the previous steps, thus eliminating any potential for cross-reactivity. The formulation of this "Double Stain Block" is a trade secret of DAKO. We guess that this elution technique is borrowed from protein Western Blotting methodology, whereby an intermediary elution buffer is also utilized that removes the previous pairs of primary/secondary antibodies (Harlow and Lane 1999).

Guided by the philosophy that if you mix your own reagent solution you know what you have got, the group of (Pirici, Mogoanta et al. 2009) tested the applicability of various elution buffers, in order to interfere with the non-covalent binding of the antibodies to their epitopes. These authors found that the blocking step with glycine SDS pH 2 buffer (25 mM glycine-HCl, 10% SDS, pH 2) for 30 min at 50°C ensures a reliable enzymatic staining without cross-reactivity and without loss of tissue antigenicity. A very important point in their study is that the authors demonstrated that most AP and HRP detection substrates, except Vector-Blue and Vector-VIP, are stable in glycine SDS pH 2 buffer, maintaining their properties even after multiple elution times. Moreover, the glycine SDS elution buffer provided an efficient inhibition of AP and HRP; this might extend its applicability to also inhibiting endogenous enzymatic activity on difficult tissues, instead of separate incubations with hydrogen peroxide or levamisole. The first attempt to perform consecutive immunostaining in the same tissue section with the use of an elution buffer was undertaken as early as 40 years ago by Tramu et al. (1978).

Concluding this section, we have to admit that *sequential* immunoenzymatic double or multiple staining severely increases the time required for staining antigens. Therefore, in our practice we prefer *simultaneous* immunoenzymatic double staining with primary antibodies noncovalently labeled with different reporter molecules using monovalent Fab fragments as a bridge. Such in-house-made Fab fragment complexes of different primary antibodies of the same species and isotype bearing different labels can be used pairwise to label multiple antigens in the same sample in a single incubation, in the absence of erroneous cross-reactions between different primary and secondary antibody pairs (see Sect. 2.2).

7.4 Nuclear Counterstaining Following Immunoenzyme Labeling

Nuclear counterstains are recommended for a better interpretation both of immunostaining and tissue morphology. The most common nuclear stain is Mayer's hematoxylin. This staining provides blue–violet nuclear staining without obscuring antigen-specific chromogen deposition, giving excellent color contrast with most commonly-used peroxidase and alkaline phosphatase substrates in brightfield microscopy (see Sect. 2.3).

Nuclear counterstaining with Mayer's hematoxylin
1. Following immunohistochemical staining, wash briefly in distilled water.
2. Stain in Mayer's hematoxylin solution for 30 sec–10 min (usually 1 min). Ready-to-use Mayer's hematoxylin solution with stabilizers purchasable from Sigma (Cat. No. MHS-16) can be stored for years without losing its potency.
3. Wash in running tap water for 5–10 min. If the running tap water is not sufficiently alkaline to blue the sections, add a few drops of ammonium hydroxide or a small pinch of calcium hydroxide to about 500 ml of water, then rinse in tap water again.
4. Rinse in distilled water and mount with an aqueous mounting medium, or proceed further with steps 5, 6 and 7.
5. Dehydrate through 95% alcohol, two changes of absolute alcohol, 5 min each.
6. Clear in two changes of xylene, 5 min each.
7. Mount with xylene-based mounting medium, like Canada balsam or DPX; however, aqueous medium can also be used (see Sect. 3.5).

Convenient ready-to-use nuclear counterstains in blue, green or red colour like Hematoxylin, Methyl Green and Nuclear Fast Red are available from Vector Laboratories (http://www.vectorlabs.com/), DAKO (http://www.dako.com/) and some other vendors, but if you are interested in preparing nuclear counterstains yourself, good and thorough procedures may be found in textbooks of Polak and Van Noorden (1997) and Renshaw (2007) or on the website: http://www.ihcworld.com/counterstain_solutions.htm.

References

Harlow E, Lane D (1999) Using Antibodies: A Laboratory Manual. Cold Spring Harbor Laboratory Press, Cold Spring Harbor, NY
Pirici D, Mogoanta L, Kumar-Singh S, Pirici I, Margaritescu C, Simionescu C, Stanescu R (2009) Antibody elution method for multiple immunohistochemistry on primary antibodies raised in the same species and of the same subtype. J Histochem Cytochem 57(6):567–575
Polak JM, Van Noorden S (1997) Introduction to immunocytochemistry. BIOS Scientific, Oxford

Renshaw S (2007) Immunohistochemistry. Scion Publishing, Cambridge
Tramu G, Pillez A, Leonardelli J (1978) An efficient method of antibody elution for the successive
    or simultaneous localization of two antigens by immunocytochemistry. J Histochem Cytochem
    26:322–324

Chapter 8
Multiple Immunofluorescence Staining

Whereas multicolor immunoenzyme staining is applicable only for separately located antigens (see Chap. 7), multicolor fluorescence immmunostaining makes it possible to colocalize antigens not only in the same cell but also in the same cellular compartment. Simultaneous immunolocalization of antigens using fluorescent antibodies can be fulfilled both by the direct (see Sect. 4.1) and indirect (see Sect. 4.2) methods. With the direct method, primary antibodies are labeled with fluorescent dyes, while with the indirect method, primary antibodies are applied as unlabeled antibodies and the visualization is performed with secondary antibodies that are labeled with fluorescent dyes.

Indirect methods for immunofluorescent detection of multiple tissue antigens in their simplest form make use of primary antibodies that are raised in different species and accordingly can be visualized with differently labeled species-specific secondary antibodies (see Sect. 8.1). However, quite often the appropriate combination of primary antibodies from different host species is not available. A general problem relates to the fact that the available primary antibodies may originate only from one species — either rabbit or mouse. When primary antibodies are raised in the same host species, the secondary species-specific antibodies can cross-react with each of the primary antibodies (Ino 2004).

However, if primary antibodies of the same species belong to different IgG isotypes, they can be selectively detected with secondary antibodies directed against the corresponding isotype (see Sect. 8.2). If the primary antibodies are of the same species and of the same IgG isotype, they can be modified via their haptenylation. Haptenylated primary antibodies are subsequently visualized with the use of secondary antibodies recognizing the corresponding hapten (see Sect. 8.3).

The cross-reaction of secondary species-specific antibodies with primary antibodies from the same species is obviously avoided by direct (one antibody layer) methods. The direct method offers an easy way for simultaneous labeling of a pair or more antigens, even when using primary antibodies from the same species. Recently, a direct technique with primary antibodies that are covalently labeled by different fluorophores was described for a simultaneous detection of up to seven

I.B. Buchwalow and W. Böcker, *Immunohistochemistry: Basics and Methods*,
DOI 10.1007/978-3-642-04609-4_8, © Springer-Verlag Berlin Heidelberg 2010

antigens (Tsurui et al. 2000). The advantages of directly labeled antibodies are many, including elimination of secondary reagents, lower backgrounds, fewer washes, less hands-on time, and improved data quality. Unfortunately, the desired fluorophore-labeled primary antibodies are not always available. Moreover, this technique suffers from a lower sensitivity compared with indirect methods. For these reasons the majority of researchers rely on indirect (two or more antibodies layer) methods for their multilabeling requirements.

8.1   Double Immunofluorescence Indirect Staining Using Primary Antibodies Raised in Two Different Host Species

Indirect methods for immunofluorescent detection of multiple tissue antigens in their simplest form make use of primary antibodies that are raised in different species. Bound primary antibodies can accordingly be visualized with differently labeled species-specific secondary antibodies. Primary antibodies to a pair of different antigens (e.g., rabbit anti-antigen A and mouse anti-antigen B) can be mixed and applied as a cocktail. The same is valid for a mix of secondary antibodies (e.g., goat anti-rabbit-FITC and goat anti-mouse-Cy3) as shown in Fig. 8.1.

Basic protocol for double immunofluorescence staining using primary antibodies raised in two different host species

(1) Deparaffinize and rehydrate tissue sections. Rinse in distilled water for 5 min.

(2) *Antigen retrieval*: place sections in a Coplin jar with antigen retrieval solution of choice (e.g., 10 mM citrate acid, pH 6) and heat at 90–110°C (depending on

Bound 1stAB from different species are selectively detected with 2nd AB directed against Ig of the corresponding species

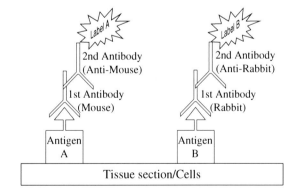

Fig. 8.1 Indirect double immunofluorescence staining using primary antibodies raised in two different host species

tissue) in a microwave, steamer, domestic pressure cooker or autoclave (see Sect. 6.1.1).

(3) *Washes and dilutions*: wash sections in PBS for 2 × 3 min. PBS is used for all washes and dilutions. Other buffers such as Tris buffered saline (TBS) may also be used.

(4) *Blocking step*: incubate sections for 20 min with PBS containing 5% normal serum of species in which the secondary antibodies were raised.

(5) *Primary antibodies*: blot excess blocking solution from sections and incubate for 60 min at room temperature or overnight at +4°C with a mixture of correspondingly diluted unlabeled primary antibodies raised in two different host species (e.g., mouse and rabbit). When using fluorophore-labeled primary antibodies as in direct immunostaining method (one antibody layer), you may skip step (6) with secondary antibodies for indirect immunostaining method (two antibodies layers). Wash sections in PBS for 3 × 3 min.

(6) *Secondary antibodies*: incubate sections for 60 min at room temperature with a pair of secondary antibodies* bearing different fluorophore labels** and raised (e.g., in goat) against the corresponding species IgG of the primary antibodies. Wash sections in PBS for 3 × 3 min.

(7) *Counterstaining*: counterstain nuclei if necessary, e.g., with DAPI (5 µg/ml PBS for 15–30 s). Wash sections in PBS for 2 × 3 min.

(8) *Optional*: to block the detachment of the fluorescent label from the antibody and to preserve the staining pattern for a longer storage, a short treatment (1–3 min) with 4% formaldehyde in PBS before mounting in water-soluble media is recommended. Wash sections in PBS for 2 × 3 min.

(9) *Mounting*: mount sections in antifade medium for fluorescence microscopy (see Sect. 3.2.2).

(10) All incubations are at room temperature unless otherwise noted.
Notes: *In general, both secondary antibodies should come from the same host species for double/multiple labeling.
**When using biotin-labeled secondary antibody, you have first to visualize biotin with a fluorophore-labeled (strept)avidin employing ABC technique (see Sect. 6.2.1), before proceeding to counterstaining (step 7).

As stated above, this approach is applicable only when primary antibodies are raised in different species. For double or multiple indirect immunofluorescence staining with primary antibodies raised in the same species, see Sects. 8.2. and 8.3 below.

8.2 Using Primary Antibodies of Different IgG Isotype

When primary mouse monoclonal antibodies are of different IgG isotypes/ subclasses, they can be selectively detected with secondary antibodies directed against the corresponding IgG isotype (Fig. 8.2).

This approach has long been used in flow cytometry. The first effort to use secondary antibodies selectively recognizing the corresponding isotype for double

Bound 1st AB from different IgG isotypes are detected with
2nd AB directed against the corresponding IgG isptype

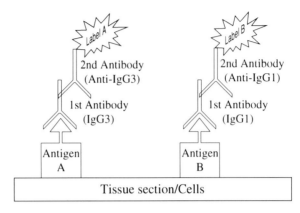

Fig. 8.2 Schematic presentation of double immunostaining using secondary antibodies (AB) raised against corresponding IgG isotypes

labeling in immunohistology originated more than 20 years ago (Tidman et al. 1981). Nowadays, secondary antibodies specifically recognizing different IgG subclasses of primary antibodies are commercially available from different manufacturers such as Abcam (Cambridge, UK), Dianova (Hamburg, Germany), BD PharMingen (Hamburg, Germany), DakoCytomation (Ely, UK), Molecular Probes (Leiden, The Netherlands), etc. With a broader availability of reliable specific anti-IgG subclass antibodies, this approach acquires a growing importance in research and clinical practice. To this end we have elaborated a basic protocol (see below) for multiple immunolabeling using monoclonal primary antibodies of different IgG isotypes (Buchwalow et al. 2005). This protocol can be widely applicable, and offers a simple procedure for simultaneous detecting two or more antigens. Here, we illustrate the application of this technique with double immunolabeling of two cytokeratins, Ck14 (IgG3) and Ck8/18/19 (IgG1), in a duct wall of the human mammary gland (Fig. 8.3).

Basic protocol for multiple immunostaining using monoclonal primary antibodies of different IgG isotypes

(1) Deparaffinize and rehydrate tissue sections. Rinse in distilled water for 5 min.
(2) *Antigen retrieval*: place sections in a Coplin jar with antigen retrieval solution of choice (e.g., 10 mM citrate acid, pH 6) and heat at 90–110°C (depending on tissue) in a microwave, steamer, domestic pressure cooker or autoclave.
(3) *Washes and dilutions*: wash sections in PBS for 2 × 3 min. PBS is used for all washes and dilutions. Other buffers such as Tris buffered saline (TBS) may also be used.
(4) *Blocking step*: incubate sections for 20 min with PBS containing 5% normal serum of species in which the secondary antibodies were raised.

(5) *Primary antibodies*: blot excess blocking solution from sections, and incubate for 60 min at room temperature or overnight at +4°C with a mixture of correspondingly diluted primary antibodies of different IgG isotypes. Wash sections in PBS for 3 × 3 min.

(6) *Secondary antibodies*: Incubate sections for 60 min at room temperature with a mixture of secondary fluorophore-conjugated antibodies raised against the corresponding IgG isotypes of primary antibodies. Wash sections in PBS for 2 × 3 min.

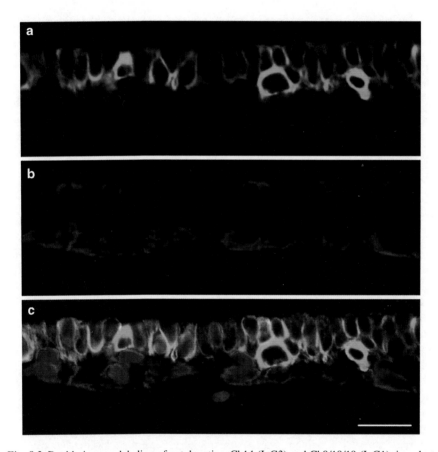

Fig. 8.3 Double immunolabeling of cytokeratins, Ck14 (IgG3) and Ck8/18/19 (IgG1), in a duct wall of the human mammary gland. Tissue section was probed simultaneously with a mixture of anti-Ck14 (IgG3) and anti-Ck8/18/19 (IgG1) mouse monoclonal primary antibodies at 1 μg/ml each, and detected after incubation with a mixture of FITC-anti-mouse IgG1 and biotin-anti-mouse IgG3 rat monoclonal antibodies with following visualization of biotin via Cy3-conjuageted Streptavidin. Nuclei were counterstained with DAPI. Separate color images for Ck8/18/19 (**a**, FITC, *green*), Ck14 (**b**, Alexa-647, pseudo-colored in *pink*) and for nuclei (DAPI, *blue*) were merged into a composite multicolor image (**c**), whereby the colocalization of both antigens was revealed through the coincidence of the two labels resulting in a hybrid color. 20-μm scale bar for entire layout. Reproduced from Buchwalow et al. (2005)

(7) *Optional*: to block the detachment of the fluorescent label from the antibody and to preserve the staining pattern for a longer storage, a short treatment (1–3 min) with 4% formaldehyde in PBS before mounting in water-soluble media is recommended. Wash sections in PBS for 2 × 3 min.

(8) *Counterstaining*: counterstain nuclei if necessary, e.g., with DAPI (for 15–30 s, 5 μg/ml PBS). Wash sections in PBS for 2 × 3 min.

(9) *Mounting*: mount sections in antifade medium for fluorescence microscopy.

(10) All incubations are at room temperature unless otherwise noted.

In its pure form, this approach is applicable to monoclonal primary antibodies belonging to different IgG subclasses. However, combined with haptenylated monoclonal antibodies or with antibodies from other host species, the limits of this technique can be widely extended. We exemplify this combined approach with a simultaneous localization of four different antigens, Ck14 (Mouse IgG3), Ck8/18/19 (Mouse IgG1), smooth muscle actin (Mouse IgG2a) and Ki67 (Rabbit polyclonal), in a section of a papillary lesion of the human mammary gland (Fig. 8.4).

8.3   Using Haptenylated Primary Antibodies

Another approach to this persistent problem relies on haptenylation of primary antibodies. Hapten (e.g., biotin, digoxigenin or any fluorophore) can be covalently bound to the antibody via *N*-hydroxysuccinimide esters (NHS-ES) (see Sect. 2.1), or conjugated employing monovalent IgG Fc-specific Fab fragments (see Sect. 2.2). Haptenylated primary antibodies can be subsequently visualized with the use of secondary antibodies recognizing the corresponding hapten (Fig. 8.5). Fluorophore-labeled primary antibodies can be directly visualized in a fluorescent microscope.

Although the desired haptenylated primary antibodies are not always available, recent developments in the technology of covalent and noncovalent labeling of antibodies allow the researcher to overcome this problem using his own in-house-made antibody conjugates (see Chap. 2). The most widely applied principle is haptenylation of amino groups via NHS-ES. For convenient protein-labeling procedures, a number of haptens (various fluorophores, enzymes, biotin, digoxigenin, etc.) are now commercially available as activated NHS-ES (Pitt et al. 1998) from various vendors (Sigma, Molecular Probes, Pierce, and many others). Innova Biosciences (http://www.innovabiosciences.com/) offers a broad choice of conjugation systems for binding fluorophore labels to antibodies. Suitable biotinylation kits (van der Loos and Gobel 2000) have been developed commercially by DakoCytomation (ARK; DakoCytomation, Ely, UK) and Molecular Probes (Zenon; Eugene, OR). In our laboratory, we prefer to carry out the noncovalent labeling of primary antibodies ourselves using labeled monovalent Fab fragments (see Sect. 2.2).

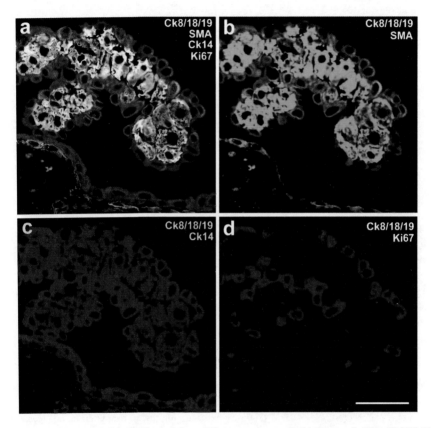

Fig. 8.4 Simultaneous localization of four different antigens, Ck14 (Mouse IgG3), Ck8/18/19 (Mouse IgG1), smooth muscle actin (SMA, Mouse IgG2a) and Ki67 (Rabbit polyclonal), in a section of a papillary lesion of the human mammary gland. Separate color images for immuno-labeling Ki-67 (Cy3, *red*), Ck8/18/19 (Alexa-350, *blue*), SMA (FITC, *green*) and Ck14 (Alexa-647, *pink*) were merged into a composite multicolor image (**a**) or double-color images designed to elucidate relations between the expression of glandular Ck8/18/19 vs SMA (**b**), glandular Ck8/18/19 vs basal Ck14 (**c**) and glandular Ck8/18/19 vs Ki67 (**d**). 50-μm scale bar for entire layout. Reproduced from Buchwalow et al. (2005)

The protocol for double/multiple immunolabeling using haptenylated primary antibodies is essentially the same as with primary antibodies of different IgG isotypes. These protocols can be easily customized depending on the availability of primary antibodies for your research requirements. For instance, you may have at your disposal a pair of monoclonal antibodies of the same IgG isotype, and only one of them is haptenylated. In this case, you have to carry out the immunostaining in two steps: in the first step you visualize the unlabeled first primary antibody with a secondary species-specific antibody, and in the second step you can detect the second primary haptenylated antibody via another secondary antibody directed against the corresponding hapten. Should the hapten be a fluorophore, it can be visualized directly in a fluorescent microscope and you do not need the second step

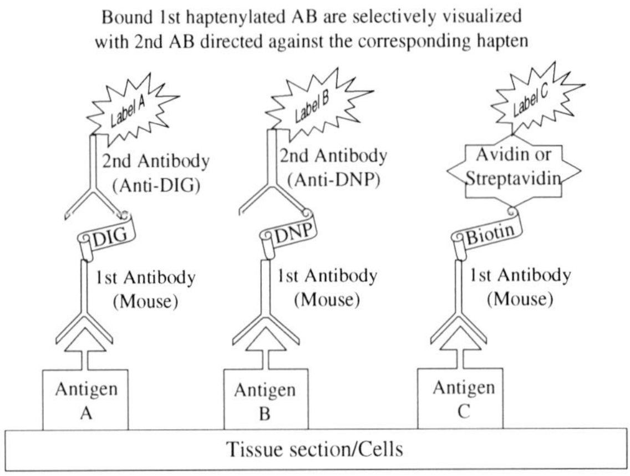

Fig. 8.5 Schematic presentation of indirect multiple immunofluorescence staining using haptenylated primary antibodies (AB) raised in the same host species

of the immunostaining with the secondary antibody. If the hapten is biotin, you can visualize the biotin label through a fluorophore-conjugated avidin or streptavidin employing the ABC technique (see Sect. 6.2.1).

References

Buchwalow IB, Minin EA, Böcker W (2005) A multicolor fluorescence immunostaining technique for simultaneous antigen targeting. Acta Histochem 107:143–148

Ino H (2004) Application of antigen retrieval by heating for double-label fluorescent immunohistochemistry with identical species-derived primary antibodies. J Histochem Cytochem 52:1209–1217

Pitt JC, Lindemeier J, Habbes HW, Veh RW (1998) Haptenylation of antibodies during affinity purification: a novel and convenient procedure to obtain labeled antibodies for quantification and double labeling. Histochem Cell Biol 110:311–322

Tidman N, Janossy G, Bodger M, Granger S, Kung PC, Goldstein G (1981) Delineation of human thymocyte differentiation pathways utilizing double-staining techniques with monoclonal-antibodies. Clin Exp Immunol 45:457–467

Tsurui H, Nishimura H, Hattori S, Hirose S, Okumura K, Shirai T (2000) Seven-colour fluorescence imaging of tissue samples based on Fourier spectroscopy and singular value decomposition. J Histochem Cytochem 48:653–662

van der Loos CM, Gobel H (2000) The animal research Kit (ARK) can be used in a multistep double staining method for human tissue specimens. J Histochem Cytochem 48:1431–1437

Chapter 9
Antigen Detection on Tissues Using Primary Antibody Raised in the Same Species

The use of a primary antibody raised in the same species as the tissue under study normally produces false immunostaining due to cross-reactivity with interstitial immunoglobulins, B cells, plasma cells and immunoglobulins bound to Fc receptors of macrophages. Antigen detection with the use of antibodies on homologous tissues (e.g., mouse antibody on mouse tissue) is complicated by severe background staining, due to the fact that secondary antibody binds to primary antibody as well as to irrelevant endogenous tissue immunoglobulins.

More often than not, antibodies with the required specificities are available in only one species. When working with mouse monoclonal antibodies on mouse tissues, a cheap alternative is to stain with antimouse isotype-specific secondary antibodies, exploiting the relative lower levels of each IgG isotype in the serum and in some cells in the mouse tissue. In this method, anti H+L pan-reactive secondary antibody must be substituted with an anti-isotype specific secondary. IgG isotypes account each for 20–25% of total serum Ig. Most mouse monoclonal antibodies working on mouse tissue are of IgG1 isotype. Be aware of possible odd cross-reactivities (e.g., many goat antimouse IgG1 heavy chain do stain rat IgG2a and IgG2b). Important, young laboratory mice raised in sterile, pathogen-free barriers, and unimmunized, have lower endogenous immunoglobulin levels than older animals raised in conventional housing (Cattoretti and Qing 2000). Unfortunately, this approach does not always help. There are, however, better ways of avoiding the cross-reactivity with endogenous tissue immunoglobulins. First, you can use haptenylated primary antibodies, which makes it possible to evade application of secondary anti-IgG antibodies that can bind both to the primary antibody and to irrelevant endogenous tissue immunoglobulins homologous to the primary antibody. Second, you can preincubate your specimen with unconjugated monovalent Fab fragments, which makes it possible to block endogenous tissue immunoglobulins that are responsible for severe background staining.

I.B. Buchwalow and W. Böcker, *Immunohistochemistry: Basics and Methods*,
DOI 10.1007/978-3-642-04609-4_9, © Springer-Verlag Berlin Heidelberg 2010

9.1   Haptenylation of Primary Antibodies with the Following Use of Secondary Antibodies Recognizing the Corresponding Hapten

In order to evade application of secondary antibodies that can bind both to the primary antibody and to irrelevant endogenous tissue immunoglobulins homologous to the primary antibody, you may use haptenylated primary antibodies. Haptenylated primary antibodies can be visualized using secondary antibodies recognizing the corresponding hapten. Primary antibodies haptenylated with a fluorophore can be directly visualized in a fluorescent microscope. Hapten (e.g., biotin, digoxigenin, enzyme or any fluorophore) can be bound to the primary antibody covalently via N-hydroxysuccinimide esters (NHS-ES) (see Sect. 2.1) or noncovalently using monovalent Fab fragments (see Sect. 2.2).

Noncovalent labeling of primary antibodies in vitro with a reporter molecule using monovalent Fab fragments that recognize both the Fc and $F(ab')_2$ regions of IgG (Brown et al. 2004) offers an easy and fast method of antibody labeling for background-free immunostaining of homologous tissue specimens. Simple mixing of labeled monovalent Fab fragments with an unconjugated primary antibody rapidly and quantitatively forms a labeling complex (labeled Fab fragment attached to primary antibody). After absorption of unbound monovalent Fab fragments with excess serum from the same species as the primary antibody, the resultant complexes can be used for immunostaining, avoiding cross-reaction of the secondary antibody with endogenous immunoglobulins when applied to tissue samples homologous to the primary antibody species. The same principle is used in the Animal Research Kit (ARK) developed by DAKO (http://www.dakousa.com/) and in Zenon Labeling Kits from Molecular Probes (http://probes.invitrogen.com/products/zenon/). However if you buy all components from Jackson Immuno-Research Laboratories Inc (http://www.jacksonimmuno.com/) to develop such in-house labeling with monovalent Fab fragments, you save a lot of money.

Procedure for generating and using primary mouse antibody–Fab fragment complexes on mouse tissues (Adapted from Brown et al. 2004)

1. Incubate specific mouse monoclonal primary antibody with FITC[*]-conjugated mouse IgG specific Fab fragments at a ratio[**] of 1:2 (weight for weight, based on concentration data supplied by manufacturers) in a small volume (e.g., in 10 µl or more, typically 1 µg of primary antibody in 10 µl) of staining buffer in a microcentrifuge tube for 20–30 min at room temperature.
2. Dilute the resultant complexes in staining buffer containing 10% normal mouse serum to give the working concentration of ≈5 µg/ml of primary mouse IgG, and incubate for 10–20 min at room temperature to block unbound Fab fragment paratopes.
3. Centrifuge complexes at 13,000 × g for 10 min, to remove insoluble precipitates (optional).

4. Further dilute (if required) the resultant primary antibody–Fab fragment complexes to optimal working concentration (usually about 1–5 µg/ml) in staining buffer containing 10–20% normal serum.

5. After preincubation of mouse tissue sections with 10% normal mouse serum for 15 min at room temperature, apply mouse antibody–Fab fragment complexes for 30–60 min at room temperature and proceed further with your standard immunostaining protocol.

Notes: *The label may also be biotin or any fluorophore. For antibody biotinylation, we use biotin-SP-conjugated AffiniPure Fab Fragment Goat Antimouse IgG (H+L) (Jackson ImmunoResearch Labs, Code Number: 115-067-003).

**The ratio of primary antibodies to Fab fragments required for the formation of complexes that produce optimal immunolabeling of specific antigen does not appear to vary dramatically with primary antibody specificity or species. Primary antibody to Fab fragment ratios of 1:2–1:4 (weight for weight, based on concentration data supplied by manufacturers) typically produces optimal results with primary mouse monoclonal antibodies.

9.2 Blocking Endogenous Tissue Immunoglobulins Homologous to Primary Antibody by Preincubation with Unconjugated Fab Fragments

An alternative approach utilizes preincubation with unlabeled monovalent Fab fragments of IgG directed against immunoglobulins of the species under study. The working concentration of unlabeled monovalent Fab fragments should not be lower than 0.1 mg/ml in PBS. This simple procedure makes it possible to block endogenous tissue immunoglobulins with practically complete elimination of background staining (Fig. 9.1).

Please note that whole IgG or a divalent $F(ab')_2$ fragment should not be used for blocking, since it has two binding sites. After blocking, some of the binding sites may remain open to capture the primary antibody introduced in a subsequent step, which may result in cross-labeling and thus create higher background.

Procedure for blocking immunoglobulins homologous to primary antibody (e.g., Mouse-on-Mouse*)

1. Bring paraffin or frozen sections to water.
2. Perform antigen retrieval if needed.
3. Wash 2 × 2 min with PBS.
4. Incubate mouse tissue sections with unconjugated AffiniPure Fab Fragment Goat Antimouse IgG (H+L) (Jackson ImmunoResearch Labs, Code Number: 115-007-003) for 1 h at room temperature. Note: (1) dilute the antibody with PBS and make sure its final concentration is not lower than 0.1 mg/ml or 100 µg/ml in PBS, (2) the Fab Fragment concentration lower than 0.1 mg/ml starts to

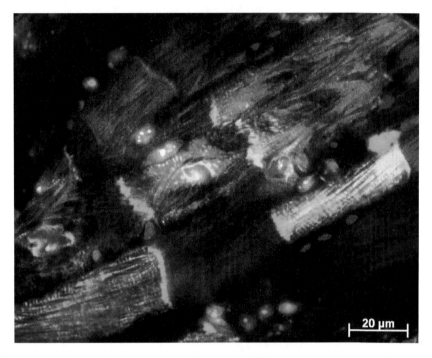

Fig. 9.1 Immunolocalization of Adenosine Receptor A (FITC, *green*) in mouse myocardium using mouse monoclonal primary antibody after blocking mouse endogenous immunoglobulins by preincubation with unconjugated Fab fragment Goat Antimouse IgG. *Red* color accounts for cardiomyocytes and erythrocytes autofluorescence captured under illumination with a filter exciting the autofluorescence in red spectrum. Nuclei are counterstained with DAPI (*blue*). Courtesy of Stephanie Grote

produce background staining, (3) concentration higher than 0.1 mg/ml may be more effective but may not be necessary, and (4) increasing blocking time to 2 h at room temperature or overnight at 4 C may be more effective. In this case, a lower concentration may be used.

5. Wash 3 × 3 min with PBS.
6. Incubate sections with primary mouse monoclonal antibody* at its optimal dilution for 30 min at room temperature.
7. Wash 2 × 2 min with PBS.
8. Incubate sections with 3% H_2O_2 in PBS for 10 min to block endogenous peroxidase.
9. Wash 3 × 2 min with PBS.
10. Incubate sections with Biotin-SP-conjugated AffiniPure Fab Fragment Goat Antimouse IgG (H+L) (Jackson ImmunoResearch Labs, Code Number: 115-067-003, 1:500 dilution) for 20–30 min at room temperature. Note: (1) longer incubation time may produce more background staining, and (2) using whole IgG secondary antibody is also possible but may bind to Fc receptors presented in some tissue types.

11. Wash 3 × 3 min with PBS.
12. Incubate sections with HRP-Streptavidin[**] (1:500, Vector Labs) in PBS for 20–30 min at room temperature.
13. Wash 3 × 3 min with PBS.
14. Incubate sections with DAB substrate solution for 5 min.
15. Wash with tap water briefly.
16. Counterstain with Hematoxylin if desired.
17. Rinse with tap water.
18. Dehydrate through 95 and 100% ethanol.
19. Clear with xylene.
20. Coverslip with permanent mounting medium.

Notes: [*]This procedure, based on the protocol from the IHC World website http://www.ihcworld.com/, is also applicable to polyclonal antibodies (e.g., on rabbit or goat tissues).

[**]For fluorescent visualization, you may use any fluorophore-conjugated Streptavidin with a subsequent nuclear counterstaining (e.g., with DAPI if needed) and mounting in any antifade mounting medium (e.g., Vectashield from Vector Laboratories, Burlingame, USA). In this case, blocking of the endogenous enzyme activity (Step 8) can be avoided.

Based apparently on the same principle as described above, suitable but rather expensive kits exploiting monovalent Fab fragments to block endogenous tissue immunoglobulins have also been developed commercially, such as Mouse-on-Mouse (M.O.M.[TM]) Kits by Vector Laboratories (http://www.vectorlabs.com/). Three Vector[®] M.O.M.[TM] kits are available. These kits use the same blocking technology and biotin-avidin detection format, but offer a choice of using either an enzyme-based or fluorescent-based visualization method.

References

Brown JK, Pemberton AD, Wright SH, Miller HRP (2004) Primary antibody-Fab fragment complexes: a flexible alternative to traditional direct and indirect immunolabeling techniques. J Histochem Cytochem 52:1219–1230
Cattoretti G, Qing Fei (2000) Application of the antigen retrieval technique in experimental Pathology: from human to mouse. In: Shi SR, Gu J, Taylor CR (eds) Antigen retrieval techniques. Eaton Publishing, Natick MA, pp 165–179

Chapter 10
Probes for Staining Specific Cellular Organelles

Many research projects require the localization of certain cellular structures, especially organelles. Specific organelle markers (endoplasmic reticulum, golgi apparatus, endosomes, lysosomes, nucleus, plasma membrane, mitochondria, peroxisome and centromer) including antibodies to organelle-specific proteins are helpful to locate these organelles by immunocytochemical or histological techniques. Many of the fluorescent probes designed for selecting organelles are able to permeate or sequester within the cell membrane (and therefore are useful in living cells), while others must be installed using monoclonal antibodies with traditional immunocytochemistry techniques (http://probes.invitrogen.com/handbook). The website http://www.organell-marker.de/powered by Acris Antibodies GmbH (Herford, Germany; http://www.acris-antibodies.com/) simplifies the search for organelle markers.

Recently, OriGene Technologies introduced GFP- and RFP-tagged TrueORF (Open Reading Frame) cDNA clones, encoding organelle-specific or structure-specific proteins. The proteins are fused with a fluorescent protein and allow direct visualization of the cell organelles (see Chap. 11) without antibodies or chemicals, as well as monitoring of protein trafficking.

10.1 Nuclear Markers

10.1.1 Nuclear Fluorescent Counterstaining

Fluorescent probes for highly selective staining of nuclei are extremely useful. Fluorescent nuclear counterstaining is recommended for a better interpretation both of immunostaining and tissue morphology. Fluorescent nuclear or DNA dyes (Table 10.1) provide highly selective nuclear staining with little or no cytoplasmic labeling; they are ideal for use as counterstains in multicolor applications.

I.B. Buchwalow and W. Böcker, *Immunohistochemistry: Basics and Methods*,
DOI 10.1007/978-3-642-04609-4_10, © Springer-Verlag Berlin Heidelberg 2010

Table 10.1 Characteristics of some fluorescent dyes used for nuclear counterstaining

Fluorochromes for nuclear counterstaining	Excitation wavelength (nm)	Emission wavelength (nm)	Color
4,6-diamidino-2-phenylindole HCl (DAPI)[a]	372	456	Blue
Hoechst 33342[a]	355	465	Blue
SYTOX Green	504	523	Green
Propidium iodide (PI)	530	615	Red

[a]Excitation spectra of DAPI and Hoechst 33342 are too short for most of the lasers and mirrors that are supplied with commercially available laser scanning microscopes, although these dyes can be imaged in conventional fluorescence microscopes with Xenon or Mercury arc-discharge lamp or when using HeNe laser/UV system or multiple photon microscopy

The stained nuclei stand out in vivid contrast to other fluorescently labeled cell structures when observed by fluorescence microscopy.

DAPI and Hoechst dyes are quite water-soluble and bind externally to AT-rich base pair clusters in the minor groove of double-stranded DNA, with a dramatic increase in fluorescence intensity at 456 nm and 465 nm respectively. Both dye classes are the most popular dyes in fluorescence microscopy for use in multicolor fluorescent labeling protocols. Their relatively low-level fluorescence emission does not overwhelm signals from green- or red-fluorescent secondary antibodies or FISH probes. Propidium iodide (PI) binds to DNA via intercalation to produce orange-red fluorescence with emission maximum at 615 nm. Nuclear dyes (DAPI, Hoechst 33342 and PI), supplied as lyophilized solids, are usually reconstituted in methanol. The stock solutions (5 mg/ml) are stable for many years when stored frozen at $\leq -20°C$ and protected from light. Before use, the stock solution is further diluted in PBS to the final concentration of 5 μg/ml. SYTOX Green is available from Molecular Probes (and comes as a solution in DMSO at 1000× concentration.) Nuclear fluorescent counterstaining is carried very rapidly (for some specimens not longer than 15 s) and should be performed after immunostaining.

Some fluorescent DNA stains can also be used for chromosome counterstaining, for detection of hybridized metaphase or interphase chromosomes in fluorescence in situ hybridization assays or for identifying apoptotic cells in cell populations (http://probes.invitrogen.com/handbook/sections/0806.html). For instance, Vybrant Apoptosis Assay Kit #4 (Molecular Probes) detects apoptosis on the basis of changes that occur in the permeability of cell membranes. This kit contains ready-to-use solutions of both YO-PRO-1 and propidium iodide nucleic acid stains. YO-PRO-1 stain selectively passes through the plasma membranes of apoptotic cells and labels them with moderate green fluorescence. Necrotic cells are stained red-fluorescent with propidium iodide.

The fluorescence spectra of the monomeric cyanine nucleic acid stains family (PO-PRO-1, BO-PRO-1 and YO-PRO-1) introduced by Molecular Probes (http://probes.invitrogen.com) cover the entire visible wavelength range. These dyes may also be used with ultraviolet trans- or epi-illuminator excitation sources. The monomeric cyanine nucleic acid stains exhibit large degrees of fluorescence enhancement upon binding to DNA (or RNA) up to 1,800-fold. Consequently, the fluorescence of unbound dye is negligible under most experimental detection conditions.

10.1.2 Cell-Proliferation Nuclear Markers

Nuclear-specific proteins, such as Ki-67 (MIB-1), proliferating cell nuclear antigen (PCNA) and protein 53 have an essential role in cell proliferation and are widely accepted proliferation marker. They are present in all dividing cells of normal and tumor tissues, but absent in resting cells.

Ki-67 nuclear antigen expression is of prognostic importance in a variety of cancers. Ki67 and MIB-1 monoclonal and polyclonal antibodies are directed against different epitopes of the same proliferation-related antigen. The monoclonal MIB-1 antibody from DAKO (Code N1633, http://www.dako.co.uk/) reacts with the Ki-67 nuclear antigen (345 and 395 kD double band in immunoblotting of protein extracted from proliferating cells) associated with cell proliferation and found throughout the cell cycle (G1, S, G2, M-phases), and absent in resting cells (G0 phase). This antibody recognizes native Ki-67 antigen and recombinant fragments of the Ki-67 molecule.

PCNA is a 36 kDa molecular weight protein also known as cyclin. PCNA is expressed in the nuclei of cells during the DNA-synthesis phase of the cell cycle. In early S phase, PCNA has a very granular distribution and is absent from the nucleoli. At late S phase, PCNA is prominent in the nucleoli. PCNA is a useful marker of cells with proliferative potential and for identifying the proliferation status of tumor tissue, which is relevant to prognosis.

Protein 53 (p53) is a transcription factor that regulates the cell cycle and plays an important role in the control of normal cell proliferation. The name p53 refers to its apparent molecular mass: it runs as a 53 kDa protein on SDS-PAGE. p53 protein binds DNA and in unstressed cells its levels are kept low through a continuous degradation. The best-described functions of p53 are cell cycle arrest and apoptosis as a response to DNA damage. p53 is a tumor-suppressor gene that, when working normally, helps to stop cells becoming cancerous. In most human cancers, the p53 gene is damaged. Monoclonal antibodies directed against epitopes at different domains of p53 protein are highly useful tools to investigate the structure–function relationship of wild-type and mutant p53 proteins. Increased p53 expression is a frequent finding in malignant tumors.

10.1.3 Nuclear Envelope Markers

The nuclear envelope is composed of the nuclear membranes (inner and outer), the nuclear lamina, and the nuclear pore complexes. The inner and outer nuclear membranes are connected at the nuclear pore sites and enclose a flattened sac

(10–40 nm wide) surrounding the nucleus. This structure is called the "perinuclear space." The perinuclear space is also continuous with rough endoplasmic reticulum space. The inner membrane of the nuclear envelope lies next to a layer of thin filaments which surrounds the nucleus except at the nuclear pores. This structure is called the "nuclear lamina." The nuclear lamina is a dense (~30–100 nm thick) fibrillar network inside the nucleus that forms a molecular interface between the inner nuclear membrane and chromatin. The principal component of the nuclear lamina is represented by so-called nuclear lamins. Nuclear lamins form a meshwork that stabilizes the inner membrane of the nuclear envelope. In lesser concentrations, nuclear lamins are distributed throughout the nucleoplasm. Nuclear lamins are also known as Class V intermediate filaments (see Sect. 10.5).

Nuclear lamins were initially identified as the major components of the nuclear lamina (Fawcett 1966). Due to their position in the nucleus, lamins were originally proposed to support the nuclear envelope and provide anchorage sites for chromatin. Recently, the nuclear lamins have also been found in the nucleoplasm. Antibodies to nuclear lamins are available commercially from many vendors [see the website: http://www.antikoerper-online.de/antigen/Lamin+A+(LMNA)/]. In addition to the lamins, vertebrates express several other lamina-associated proteins including: LAP1, of which there are three isoforms (α, β and γ); LAP2 (six isoforms); emerin; MAN1; lamin B receptor (LBR); otefin; ring-finger binding protein (RFBP); and nurim (Dechat et al. 2000; Cohen et al. 2001).

10.1.4 Nucleolar Markers

There is increasing evidence that nucleoli play important roles in the regulation of many fundamental cellular processes, including cell cycle regulation, apoptosis, telomerase production, RNA processing, monitoring and response to cellular stress. These multiple functions can be achieved by the transient localization of several hundred proteins within the nucleolar structure. The abundant nucleolar proteins nucleophosmin (B23) and nucleolin (C23) have been the subject of numerous studies (Ugrinova et al. 2007 and citations therein). B23 and C23 are useful markers for tumor diagnosis and prognosis. Their higher expression correlates with cell-proliferative status and is associated with poorer disease-specific survival (Li et al. 2007). It has been suggested that B23 might be involved in the nucleolar localization of another nuclear protein p120 (Valdez et al. 1992).

C23 and B23 are argyrophilic proteins targeting to nucleolar organizing regions (NORs). They can also be identified by silver staining at low pH. NORs are defined as nucleolar components containing a set of argyrophilic proteins, which are selectively stained by silver methods. After silver staining, the NORs can be easily identified as black dots exclusively localized throughout the nucleolar area, and are called "AgNORs." The NORs' argyrophilia is due to a group of nucleolar proteins, which have a high affinity for silver (AgNOR proteins). A number of studies carried out in different tumor types demonstrated that malignant cells frequently present a greater AgNOR protein amount than corresponding non-malignant cells. Over the past years, the "AgNOR method" has been applied in tumor pathology for both diagnostic and prognostic purposes (Treré 2000).

The first silver-staining methods employed for AgNOR protein visualization consisted of two successive phases: an impregnation step with silver nitrate ($AgNO_3$) and a development step with a reducing agent (such as ammonia or formic acid) (Goodpasture and Bloom 1975). In order to control the staining reaction better and avoid non-specific background, Ploton et al. (1986) proposed a lower temperature and a longer staining time. In this method, the Ag-NOR solution is freshly prepared by dissolving gelatin at a concentration of 2 g/dl in 1 g/dl aqueous formic acid to form the first solution. This solution is combined with 50 g/dl aqueous silver nitrate solution (1:2, v/v) to give the final Ag–NOR solution. The Ag–NOR solution is then immediately poured over the deparaffinized sections, which are then left in the dark, at room temperature, for 40 min. The silver colloid is washed from the sections with distilled deionized water. Because of its reliability and specificity, this silver-staining method has become the most frequently employed for AgNOR protein visualization in routine cytohistopathology (Treré 2000).

There are a number of other proliferation-associated proteins targeted to the nucleolus, such as protein p120 and fibrillarin. Protein p120 was originally identified in the nucleoli of human tumor cells by the use of monoclonal antibodies. Protein p120 is involved in rRNA/ribosome maturation and the level of its expression increases during cell proliferation. Another nucleolar marker, fibrillarin, also known as FBL, is a component of a nucleolar small nuclear ribonucleoprotein (snRNP) particle. It is thought to participate in the first step in processing preribosomal (r)RNA. It is associated with the U3, U8, and U13 small nuclear RNAs and is located in the dense fibrillar component of the nucleolus. Antibody to FBL can be purchased from Santa Cruz Biotechnology, inc. (http://www.scbt.com/) and from Abcam, Cambridge, UK (http://www.abcam.com/). This antibody detects a band at close to 34 kDa in all species tested.

10.2 Probes for Mitochondria

Mitochondria are distinct organelles with two membranes. The outer membrane limits the organelle and the inner membrane is thrown into folds or shelves that project inward and are called "cristae mitochondriales." The uptake of most mitochondrion-selective dyes is dependent on the mitochondrial membrane potential. Conventional fluorescent stains for mitochondria, such as rhodamine and tetramethylrosamine, are readily sequestered by functioning mitochondria. They are, however, subsequently washed out of the cells once the mitochondrion's membrane potential is lost. This characteristic limits their use in experiments in which cells must be treated with aldehyde-based fixatives or other agents that affect the energetic state of the mitochondria. To overcome this limitation, the research team of Molecular Probes (Poot et al. 1996) has developed MitoTracker probes — a series of patented mitochondrion-selective stains that are concentrated by active mitochondria and well-retained during cell fixation (available from Molecular Probes, http://probes.invitrogen.com). Because the MitoTracker Orange, Mito-Tracker Red and MitoTracker Deep Red probes are also retained following permeabilization; the sample retains the fluorescent staining pattern characteristic of live cells during subsequent processing steps for immunocytochemistry, in situ hybridization or electron microscopy. To label mitochondria, cells are simply incubated in submicromolar concentrations of the MitoTracker probe, which passively diffuses across the plasma membrane and accumulates in active mitochondria. Once the mitochondria are labeled, the cells can be treated with aldehyde-based fixatives to allow further processing of the sample. MitoTracker probes are cell-permeant mitochondrion-selective dyes that contain a mildly thiol-reactive chloromethyl moiety. The chloromethyl group appears to be responsible for keeping the dye associated with the mitochondria after fixation. In addition, MitoTracker reagents eliminate some of the difficulties of working with pathogenic cells because, once the mitochondria are stained, the cells can be treated with fixatives before the sample is analyzed.

The mouse monoclonal antibodies to mitochondria from Abcam (http://www.abcam.com/index.html?datasheet=3298), from Acris (http://www.acris-antibodies.com/products/acris/BM608/antibody/Mitochondria.html) and from some other vendors, show specific reactivity to a 60 kD mitochondrial antigen in all human cell types. Fresh cells cultured on microscope slide are recommended, while formalin-fixed paraffin embedded tissue sections can also be suitable. Use muscle tissue for positive control. To find mitochondria antibodies from different companies by species reactivity, application, host species, and/or conjugate, use the website "Biocompare Buyer's Guide": http://www.biocompare.com/matrixsc/3194/2/6/26933/Mitochondria.html. Alternatively, mitochondria can be visualized either with specific antibodies raised against mitochondrial proteins, such as porin or pyruvate dehydrogenase, or with antibodies against various subunits of the oxidative phosphorylation complex, for instance, cytochrome oxidase. Mitochondrial subcellular localization is also possible using vital staining with a cationic dye Janus green B or employing enzyme histochemical staining for peroxidase or cytochromoxidase (Lojda et al. 1976; Polak and Van Noorden 1997).

10.3   Probes for Endoplasmic Reticulum and Golgi Apparatus

Several antibodies specifically targeting to Golgi complex or endoplasmic reticulum are available commercially from many vendors. The SelectFX Alexa Fluor 488 Endoplasmic Reticulum Labeling Kit from Molecular Probes (http://www.invitrogen.com/site/us/en/home/Applications.html) provides all the reagents required to fix and permeabilize mammalian cells and then specifically label the endoplasmic reticulum. This kit employs a primary antibody (Mouse IgG2b) directed against an endoplasmic reticulum-associated protein, protein disulfide isomerase (PDI), and an Alexa Fluor 488 dye-labeled secondary antibody.

For identification of the Golgi apparatus, Molecular Probes offers anti-human golgin-97 mouse monoclonal antibody CDF4 (Mouse IgG1). This antibody recognizes a protein unique to the Golgi apparatus of most vertebrate species (however, it may not work for rat cells) and is therefore useful for immunodetection and identification of the Golgi apparatus in most cells. Anti-human golgin-97 antibody was originally isolated from the serum of a patient with the autoimmune disease known as Sjögren's syndrome. This antibody recognizes a 97 kDa protein called golgin-97, a member of the granin family of proteins and a peripheral membrane protein localized on the cytoplasmic face of the Golgi apparatus.

Monoclonal anti-Golgi 58 K Protein/Formiminotransferase Cyclodeaminase (FTCD) antibody (mouse IgG1, clone 58 K-9) is supplied by Sigma as ascites fluid (http://www.sigmaaldrich.com/sigma-aldrich/home.html). This antibody recognizes an epitope located on the microtubule-binding peripheral Golgi membrane 58 kDa protein. In addition to the Golgi apparatus, FTCD is localized to the centrosome, more abundantly around the mother centriole. The centrosome localization of FTCD continues throughout the cell cycle and is not disrupted after Golgi fragmentation induced by colcemid and brefeldin A. FTCD in the centrosome may be associated with polyglutamylated residues of centriole microtubules and may play a role in providing centrioles with glutamate produced by cyclodeaminase domains of FTCD (Hagiwara et al. 2006). Therefore it is also useful for studies on the effect of microtubule-perturbing agents on the Golgi apparatus.

Wheat germ agglutinin (WGA) conjugates can also be used as markers of the trans-Golgi (Guasch et al. 1995). The Golgi complex and Golgi-derived vesicles may be stained using lectins specific for sugar residues present in this compartment, because one of the primary functions of the Golgi is glycosylation of proteins (Virtanen et al. 1980).

10.4 Endocytic Pathways

In most cells, the major route for endocytosis is mediated by the molecule clathrin. Clathrin is a major protein component of the cytoplasmic face of intracellular organelles, called coated vesicles and coated pits. A variety of mono- and polyclonal anti-clathrin antibodies are purchasable from Santa Cruz Biotechnology, inc. (http://www.scbt.com/table-clathrin.html).

Shortly after formation of coated vesicles, the clathrin coat is removed and the vesicles are referred to as endosomes. Endosomes are roughly 300–400 nm in diameter when fully mature. Antibodies to earlier and late endosomes are available from antibodies-online GmbH, Aachen, Germany (http://www.antikoerper-online. de/). Early endosomal antigen 1 (EEA1) is a 162 kDa membrane-bound protein component specific to the early endosomes and is essential for their fusion with early endocytic vesicles for subsequent redistribution of extracellular compounds to alternate destinations. Extracellular materials trapped in the endocytic vesicles can be either passed into the endosomal compartment or returned to the surface. Some materials that reach the late endosomes are degraded in lysosomes.

Lysosomes are organelles containing digestive enzymes — acid hydrolases. They are used for the digestion of macromolecules from phagocytosis, endocytosis, and autophagy. Lysosomal hydrolases can be stained using enzyme histochemical methods (Lojda et al. 1976; Van Noorden and Frederiks 1993). The most useful tools for investigating lysosomal properties with fluorescence microscopy are the LysoTracker and LysoSensor dyes developed by Molecular Probes. These structurally diverse agents contain heterocyclic and aliphatic nitrogen moieties that modulate transport of the dyes into the lysosomes of living cells for both short-term and long-term studies.

Lysosomal markers — antibodies to LAMP-1 (also designated CD107a) and LAMP-2 (also designated CD107b) — are available from Santa Cruz Biotechnology, Inc. (http://www.scbt.com/). Lysosome-associated membrane proteins, LAMP-1 and LAMP-2 are involved in a variety of functions, including cellular adhesion, and are thought to participate in the process of tumor invasion and metastasis. LAMP-1 and LAMP-2 proteins are sorted at the trans-Golgi network and transported intracellularly via a pathway that is distinct from the clathrin-coated vesicles. Both LAMP-1 and LAMP-2 are involved in maintaining lysosome acidity and protecting the lysosomal membranes from autodigestion. Their expression is increased in patients with lysosomal storage diseases, such as mucopolysaccharidoses, GM_2 gangliosidoses, lipid storage disorders, glycoproteinoses, mucolipidoses, or leukodystrophies. These diseases are associated with the accumulation of substrates within the cell, impairing metabolism.

Caveolae (Latin for *little caves*, singular: caveola) are small (50–100 nm) invaginations of the plasma membrane. They were found in most cell types and are particularly numerous in the continuous endothelium of certain microvascular beds (e.g., heart, lung, and muscles) in which they have been originally identified as transcytotic vesicular carriers. Some cell types, such as neurons, may completely lack caveolae. Many functions are ascribed to caveolae, ranging from clathrin-independent endocytosis and transcytosis of various macromolecules (including LDL) to signal transduction (compartmentalization of certain signaling molecules). Caveolae are responsible for the regulation of important metabolic pathways including direct interaction with G-protein alpha subunits and functional regulation of their activity. Caveolae may play a critical role in several human diseases such as atherosclerosis, cancer, diabetes, and muscular dystrophies.

The principal structural component of caveolae membrane is caveolin, a 21–24-kDa integral membrane protein. The caveolin gene family is constituted by three members: CAV1, CAV2, and CAV3, coding for the proteins caveolin-1, caveolin-2 and caveolin-3, respectively. CAV1 and CAV2 are located next to each other on chromosome 7 and express colocalizing proteins that form a stable hetero-oligomeric complex. Caveolin-1 and caveolin-2 proteins are most abundantly expressed in endothelial cells, smooth muscle cells, skeletal myoblasts, fibroblasts, and 3T3-L1 cells differentiated to adipocytes. In skeletal muscle cells and cardiac myocytes, caveolae contain caveolin-3 (or M-caveolin), a muscle-specific isoform of the scaffolding protein caveolin (for review see Feron et al. 1997). A variety of mono- and polyclonal antibodies to all caveolin isoforms are purchasable from Santa Cruz Biotechnology, inc. (http://www.scbt.com/table-caveolin.html), Invitrogen (http://www.invitrogen.com/site/us/en/home.html) and Abcam (http://www.abcam.com/).

10.5 Probes for Cytoskeleton

The cytoskeleton is the collective name for all structural filaments in the cell. The cytoskeletal filaments are involved in establishing cell shape, and providing mechanical strength, locomotion, intracellular transport of organelles and chromosome separation in mitosis and meiosis. The cytoskeleton is made up of three kinds of protein filaments: actin filaments (also called microfilaments), intermediate filaments and microtubules.

Actin filaments are the thinnest of the cytoskeletal filaments, and therefore also called microfilaments. Polymerized actin monomers form long, thin fibers of about 8 nm in diameter. Along with the above-mentioned function of the cytoskeleton, actin interacts with myosin ("thick") filaments in skeletal muscle fibers to provide the force of muscular contraction. Actin/Myosin interactions also help produce cytoplasmic streaming in most cells.

Intermediate filaments average 10 nm in diameter [and thus are "intermediate" in size between actin filaments (8 nm) and microtubules (25 nm) or the thick filaments of skeletal muscle fibers]. There are several types of intermediate filament: keratins are found in epithelial cells; nuclear lamins form a meshwork that stabilizes the inner membrane of the nuclear envelope; neurofilaments strengthen the long axons of neurons; vimentins provide mechanical strength to muscle (and other) cells. Keratins are alpha-type fibrous polypeptides with a diameter of 7–11 nm. They are important components of the cytoskeleton in almost all epithelial cells as well as in some non-epithelial cell types. Keratins were earlier thought to be separable into "hard" and "soft," or "cytokeratins" and "other keratins," but these designations are now understood to be incorrect. In 2006, a new nomenclature (Schweizer et al. 2006) was adopted for describing keratins which takes this into account. Keratins are, generally, held to be the most ubiquitous markers of epithelial differentiation, and, so far, 20 distinct types numbered by Moll (Moll et al. 1982, 1992) have been revealed. For further detail about keratin immunostaining see Sect. 13.1.

Microtubules are straight, hollow cylinders about 25 nm in diameter built by the assembly of dimers of alpha tubulin and beta tubulin. They participate in a wide variety of cell activities. Most involve motion. The motion is provided by protein "motors" that use the energy of ATP to move along the microtubule. There are two major groups of microtubule motors: kinesins and dyneins.

Microtubules grow by the polymerization of tubulin dimers powered by the hydrolysis of GTP. They are commonly organized by the centrosome. The centrosome is located in the cytoplasm attached to the outside of the nucleus. As mitosis proceeds, microtubules grow out from each centrosome toward the metaphase plate. The centrosome can be specifically immunostained using anti-gamma-tubulin antibodies that can serve as centrosome marker. The abundance of gamma tubulin is less than 1% of the level of either alpha or beta tubulin. It shares approximately 28–32% identity with alpha tubulin from various organisms and 32–36% identity with beta tubulins. Antibody against gamma tubulin [GTU-88] also referred to as Centrosome Marker (ab11316) is available from Abcam.

The detection, localization and characterization of cytoskeletal proteins are fundamental to the understanding of such cell functions as: establishing cell shape, and providing mechanical strength, locomotion, intracellular transport of organelles and chromosome separation in mitosis and meiosis. Cytoskeleton, Inc. (http://www.cytoskeleton.com/) offers an extensive line of cytoskeletal, cytokeratin, and intermediate filament antibodies. For successful in vitro research on cytoskeletal components, it is vital to use the right buffers. For example, both tubulin and actin polymerization are highly dependent on the buffer conditions. Cytoskeleton, Inc. also provides a numbers of buffers and reagents (http://www. cytoskeleton.com/products.html) for cytoskeletal research. Diverse antibodies to cytoskeletal and cytoskeleton-associated proteins are also purchasable from Sigma (http://www.sigmaaldrich.com/life-science/cell-biology/antibodies/learning-center/ antibody-explorer/primary-antibodies/cytoskeleton.html), Abcam, Serotec, Dako, Biozol and other vendors.

10.6   Phalloidin Probes for Actin Filaments

Filamentous actin (F-actin) may be efficiently stained using labeled phallotoxins isolated from a deadly mushroom *Amanita phalloides*. Pioneering work on these toxins was done by the Nobel laureate Heinrich Wieland in the 1930s. The most commonly used member of this toxin family, phalloidin, may be purchased conjugated to biotin or to a wide variety of fluorescent dyes. Phalloidin selectively binds and stabilizes polymerized, filamentous actin without binding monomeric actin. This property makes phalloidin more attractive than actin-specific antibodies for visualization of filamentous actin. Phalloidin as a useful tool for investigating the distribution of F-actin in cells using fluorescently labeled phalloidin was first reported by Cooper (1987). Using phalloidin conjugated to the fluorophore eosin which acts as the fluorescent tag, Capani et al. (2001) developed a method for ultrastructural localization of F-actin. In this method, known as fluorescence photooxidation, fluorescent molecules can be utilized to drive the oxidation of diaminobenzidine (DAB) to create a reaction product that can be rendered electron-dense and detectable also by electron microscopy.

For labeling F-actin, Molecular Probes offers several fluorescent phalloidin derivatives supplied as lyophilized solids. Once reconstituted in methanol, the stock solutions (final concentration of 200 units/ml, which is equivalent to approximately 6.6 μM) are stable for at least one year when stored frozen at $\leq -20°$C, desiccated, and protected from light. Working solution is prepared immediately before use by diluting 10 μL of the methanolic stock solution into 200 μl PBS or whatever buffer is being used. Staining with biotinylated phalloidin requires the use of approximately a five-times higher concentration of the phalloidin conjugate than when staining with fluorescent phalloidin; additionally a fluorescent or enzyme-conjugated avidin or streptavidin detection reagent must be used.

Since phalloidin is cell-membrane impermeable, cells or tissue specimens must be fixed in buffered formalin before staining. Note that methanol can disrupt actin during the fixation process. Aldehyde fixation renders the plasma membrane permeable enough; therefore we have not found the additional step with detergent permeabilization to be necessary. Phalloidin-stained specimens should be imaged within a few days of staining, as the stain will dissociate into the mounting medium over time and produce a background autofluorescence that may obscure fine detail. Routinely used in our laboratory, the following protocol for F-actin labeling with fluorescent phalloidins is principally based on the manufacturer's guidelines (Molecular Probes: http://probes.invitrogen.com/media/pis/mp00354.pdf).

F-actin labeling with fluorescent phalloidins

1. Fix the sample in 4% formaldehyde* solution in PBS for 15 min.
2. Wash two or more times with PBS.
3. When staining with any of the fluorescent phalloidins, dilute 10 μL methanolic stock solution into 400 μl PBS. To reduce nonspecific background staining with these conjugates, add 1% bovine serum albumin (BSA) to the staining solution. It may also be useful to preincubate fixed cells with PBS containing 1% BSA.
4. Apply the phalloidin-staining solution for 20 min at room temperature (generally, any temperature between 4°C and 37°C is suitable).
5. Wash two or more times with PBS.
6. Counterstain nuclei if necessary, e.g., with DAPI (5 μg/ml PBS). Wash sections in PBS for 2 × 3 min.
7. Mount sections in antifade medium for fluorescence microscopy.
8. All incubations are at room temperature unless otherwise noted.
   Note: *Be aware that phalloidin labeling is incompatible with alcohol treatment.

Labeling with fluorescent phalloidins may be combined with immunostaining. In this case, the phalloidin-staining solution can be applied in a mixture with fluorescently labeled secondary antibodies. Combination of immunostaining with fluorescent phalloidins and fluorescently counterstained nuclei are extremely useful in multiple labeling strategies to locate antigens of interest with specific components of the cell.

References

Capani F, Deerinck TJ, Ellisman MH, Bushong E, Bobik M, Martone ME (2001) Phalloidin-eosin followed by photo-oxidation: a novel method for localizing F-actin at the light and electron microscopic levels. J Histochem Cytochem 49:1351–1361

Cohen M, Lee KK, Wilson KL, Gruenbaum Y (2001) Transcriptional repression, apoptosis, human disease and the functional evolution of the nuclear lamina. Trends Biochem Sci 26(1):41–47

Cooper JA (1987) Effects of cytochalasin and phalloidin on actin. J Cell Biol 105:1473–1478

Dechat T, Korbei B, Vaughan OA, Vlcek S, Hutchison CJ, Foisner R (2000) Lamina-associated polypeptide 2alpha binds intranuclear A-type lamins. J Cell Sci 113(19):3473–3484

Fawcett DW (1966) On the occurrence of a fibrous lamina on the inner aspect of the nuclear envelope in certain cells of vertebrates. Am J Anat 119:129–145

Feron O, Smith TW, Michel T, Kelly RA (1997) Dynamic targeting of the agonist-stimulated m2 muscarinic acetylcholine receptor to caveolae in cardiac myocytes. J Biol Chem 272 (28):17744–17748

Goodpasture C, Bloom SE (1975) Visualization of nucleolar organizer regions in mammalian chromosomes using silver staining. Chromosoma 20:37–50

Guasch RM, Guerri C, Oconnor JE (1995) Study of surface carbohydrates on isolated Golgi subfractions by fluorescent lectin-binding and flow-cytometry. Cytometry 19:112–118

Hagiwara H, Tajika Y, Matsuzaki T, Suzuki T, Aoki T, Takata K (2006) Localization of Golgi 58 K protein (formiminotransferase cyclodeaminase) to the centrosome. Histochem Cell Biol 126:251–259

Li J, Sejas DP, Burma S, Chen DJ, Pang Q (2007) Nucleophosmin suppresses oncogene-induced apoptosis and senescence and enhances oncogenic cooperation in cells with genomic instability. Carcinogenesis 28:1163–1170

Lojda Z, Gossrau R, Schiebler T (1976) Enzyme histochemistry. A laboratory manual. Springer, Berlin

Moll R, Franke WW, Schiller DL (1982) The catalog of human cytokeratins: patterns of expression in normal epithelia, tumors and cultured cells. Cell 31:11–24

Moll R, Löwe A, Laufer J, Franke WW (1992) Cytokeratin 20 in human carcinomas. A new histodiagnostic marker detected by monoclonal antibodies. Am J Pathol 140:427–447

Ploton D, Menager M, Jeannesson P, Himber G, Pigeon F, Adnet JJ (1986) Improvement in the staining and in the visualization of the argyrophilic proteins of the nucleolar organizer region at the optical level. Histochem J 18:5–14

Polak JM, Van Noorden S (1997) Introduction to immunocytochemistry. BIOS Scientific Publishers, Oxford

Poot M, Zhang YZ, Kramer JA, Wells KS, Jones LJ, Hanzel DK, Lugade AG, Singer VL, Haugland RP (1996) Analysis of mitochondrial morphology and function with novel fixable fluorescent stains. J Histochem Cytochem 44:1363–1372

Schweizer J, Bowden PE, Coulombe PA, Langbein L, Lane EB, Magin TM, Maltais L, Omary MB, Parry DA, Rogers MA, Wright MW (2006) New consensus nomenclature for mammalian keratins. J Cell Biol 174(2):169–174

Treré D (2000) AgNOR staining and quantification. Micron 31:127–131

Ugrinova I, Monier K, Ivaldi C, Thiry M, Storck S, Mongelard F, Bouvet P (2007) Inactivation of nucleolin leads to nucleolar disruption, cell cycle arrest and defects in centrosome duplication. BMC Mol Biol 8:66

Valdez BC, Perlaky L, Saijo Y, Henning D, Zhu C, Busch RK, Zhang WW, Busch H (1992) A region of antisense RNA from human p120 cDNA with high homology to mouse p120 cDNA inhibits NIH 3 T3 proliferation. Cancer Res 52(20):5681–5686

Van Noorden CJF, Frederiks WM (1993) Enzyme Histochemistry: A Laboratory Manual of Current Methods. Bios Scientific Publishers Ltd., Oxford

Virtanen I, Ekblom P, Laurila P (1980) Subcellular compartmentalization of saccharide moieties in cultured normal and malignant-cells. J Cell Biol 85:429–434

Chapter 11
The Use of Epitope Tags in Histochemistry

An epitope (an antigenic determinant) is a part of a molecule to which an antibody binds. If you do not have an antibody raised against the antigen, which you want to visualize, you might consider adding an epitope tag onto this molecule (usually proteins). This technique can be used to detect proteins that are very difficult to isolate and purify for antibody generation, not to say that epitope tagging can be done at a lower cost than antibody preparation. Moreover, this technique can be used to distinguish between two proteins with similar antigenicity.

Adding an epitope tag onto the molecule is done as follows: the gene from the target protein is inserted into the epitope tag vector, and the target protein with its tag is expressed in cells by transfection of the vector. A number of methods can be used to introduce the reporter gene, including pronuclear injection, retroviral-mediated gene transfer, or gene transfer in embryonic stem cells. The classic and most widely used route for the production of transgenics is through the introduction of a DNA construct linking a promoter/enhancer element to the reporter gene into zygotes by pronuclear injection. The tagged protein will be distributed inside the cell according to the original properties of the protein. For more information on obtaining optimal expression of cloned proteins in mammalian cells see, for instance, Kolodziej and Young (1991) and Chen et al. (2006).

In most cases, epitope tags are constructed of amino acids. Epitope tags range from 6 to 15 amino acids long and are designed to create a molecular handle for your protein. Although any short stretch of amino acids known to bind an antibody could become an epitope tag, there are a few that are especially popular. Three examples include: *c-myc* – a 10 amino acid segment of the human proto-oncogene myc (EQKLISEEDL); *HA* – Hemagglutinin protein from human influenza hemagglutinin protein (YPYDVPDYA); *HIS* – six histidines placed in a row can also be used as an epitope tag. When using such invisible epitope tags, the target proteins can be visualized by immunohistochemical procedures with anti-tag antibodies.

Due to the importance of this technology for protein expression, manufacturers — Sigma, Serotec, Abcam and many others — offer a wide range of products for the detection, isolation and purification of tagged proteins. These vendors also provide monoclonal and polyclonal antibodies specific for the most commonly

I.B. Buchwalow and W. Böcker, *Immunohistochemistry: Basics and Methods*,
DOI 10.1007/978-3-642-04609-4_11, © Springer-Verlag Berlin Heidelberg 2010

used epitope tags, as well as antibody–enzyme conjugates, fluorescent antibodies, antibody-affinity resins and the specific peptide tag sequence.

The use of epitope tagging is similar to the approach for constructing fusion proteins with the green fluorescent protein (GFP). When using GFP, the target protein can be observed directly because the GFP tag emits fluorescence. GFP may be visualized using existing FITC filter sets even though it fluoresces at slightly shorter wavelengths. GFP has a major excitation peak at a wavelength of 395 nm and an emission peak, at 509 nm (see Sect. 14.2.1).

The gene that codes for GFP was cloned by (Prasher et al. 1992). GFP was isolated from the jellyfish *Aequorea victoria*. Using biotechnology methods, the GFP gene is fused with a host gene of interest, and this chimera is transfected into the host genome. The resulting fusion protein that the cell produces is fluorescent. GFP has become one of the most popular reporter proteins. Along with its mutated allelic forms, blue, cyan, and yellow fluorescent proteins, GFP is used to construct fluorescent chimeric proteins that can be expressed in living cells, tissues, and entire organisms after transfection with the engineered vectors. Other fluorescent proteins, red fluorescent proteins (RFP), have been isolated from other species, including coral reef organisms, and are similarly useful. The natural fluorescence of such epitope tags allow researchers to optically detect specific types of cells both in vitro or even in vivo (in the living organism, including mammalians) using fluorescence microscopy. Commercially available anti-GFP antibodies may also be used for immunohistochemical amplification of the GFP signal.

As might be expected, denaturation of GFP can destroy fluorescence. However, formalin fixation with subsequent paraffin embedding practically does not alter the fluorescence properties of GFP. Figure 11.1 demonstrates the localization of GFP-expressing virus in a paraffin section of formaldehyde-fixed mouse myocardium. Moreover, the GFP fluorescence is retained in specimens after subsequently performed immunohistochemical labeling. Figure 11.2 demonstrates localization of GFP and calsequestrin (CSQ) in a monolayer of cultured neonatal rat cardiac myocytes. Myocytes were infected with a recombinant adenovirus containing the cDNAs of GFP and CSQ. CSQ was visualized immunohistochemically using anti-CSQ antibodies.

Alternatively, GFP can be visualized using rabbit polyclonal antibody raised against GFP purified directly from *A. victoria*. This anti-GFP antibody facilitates the detection of native GFP, recombinant GFP, and GFP-fusion proteins both by immunofluorescence and brightfield microscopy, as well as by western blot analysis and immunoprecipitation. Direct anti-GFP conjugates made from a complete serum or from an IgG fraction are available from Invitrogen (http://www.invitrogen.com/site/us/en/home.html). Additional options for your research offered by Invitrogen include two mouse monoclonal antibodies and a chicken IgY fraction.

Recently, OriGene Technologies (http://www.origene.com/) introduced GFP- and RFP-tagged TrueORF (Open Reading Frame) cDNA clones encoding organelle-specific or structure-specific proteins. The proteins are fused with a fluorescent protein and allow direct visualization of the organelles or structures (see Chap. 10).

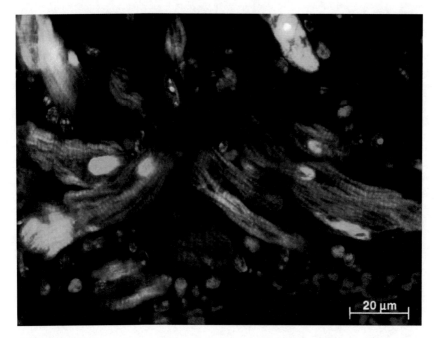

Fig. 11.1 Localization of adenoviral mediated GFP expression in the mouse myocardium. GFP is targeted to T-tubules along myofibers and to nuclei of cardiomyocytes. Red autofluorescence of cardiomyocytes and erythrocytes was captured with a filter exciting the autofluorescence in red spectrum under a longer exposure than with the filter exciting fluorescence in the green spectrum. Nuclei are counterstained with DAPI. Courtesy of Larissa Fabritz

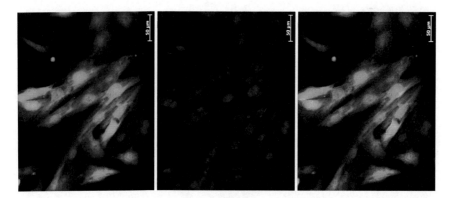

Fig. 11.2 Localization of GFP and calsequestrin (CSQ) in neonatal rat cardiac myocytes. Myocytes were infected with a recombinant adenovirus containing the cDNAs of GFP and CSQ. Both cDNAs were expressed under control of a separate cytomegalovirus promoter. Expression of CSQ and the coexpressed GFP was detected by fluorescence microscopy. *Left*: GFP fluorescence (*green*), *middle*: immunostaining of CSQ (*red*), *right*: overlay of GFP fluorescence and immunostaining. Nuclei were counterstained with DAPI (*blue*). Courtesy of Ulrich Gergs

References

Chen YIG, Maika SD, Stevens SW (2006) Epitope tagging of proteins at the native chromosomal loci of genes in mice and in cultured vertebrate cells. J Mol Biol 361:412–419

Kolodziej PA, Young RA (1991) Epitope tagging and protein surveillance. Methods Enzymol
˙ 194:508–519

Prasher DC, Eckenrode VK, Ward WW, Prendergast FG, Cormier MJ (1992) Primary structure of the *Aequorea victoria* green-fluorescent protein. Gene 111:229–233

Useful Websites

http://www.olympusfluoview.com/applications/epitopetagging.html
https://www.roche-applied-science.com
http://www.bio.davidson.edu/COURSES/GENOMICS/method/Epitopetags.html

Chapter 12
Immunohistochemistry at the Ultrastructural Level

Immunohistochemistry at the ultrastructural level, also known as immunoelectron microscopy, can be performed on ultrathin cryosections or on ultrathin sections of probes embedded in epoxy or acrylic resins. Ultrathin frozen sections are prepared from aldehyde-fixed specimens that have been infused with cryoprotectants such as sucrose and then frozen in liquid nitrogen. After freezing, the sections are cut using glass or diamond knives at between $-90°C$ and $-120°C$. They are then thawed, mounted on plastic-coated EM grids and immunolabeled. The review of the theoretical background, equipment, methods, and applications of cryofixation and cryoultramicrotomy (Saga 2005) can be found in the Internet under address: http://www.springerlink.com/content/x763346q14552851/fulltext.pdf.

Unfortunately, cryoultramicrotomy is not free from drawbacks, such as that the sections are prone to folding and difficult to prepare in series. In contrast to ultrathin cryosections, resin embedding provides the best tissue preservation, excellent structure contrast, and higher stability under the electron beam, and sample blocks are much easier to cut. With the development of effective methods of etching and heat-induced antigen retrieval (see Sect. 12.3) on ultrathin sections cut from resin-embedded tissue probes (Groos et al. 2001; Hayat 2002), ultrathin cryosections are becoming much less exploited for immunolabeling at the electron microscope level than resin sections.

12.1   Colloidal Gold Conjugates

Immunohistochemical labeling for electron microscopy is based on the same principles as immunohistochemistry for light microscopy. The differences are that specimen sections must be much thinner (50–100 nm) and the label must be electron-dense. The first electron-dense labels used for immunolabeling at the electron microscope level were ferritin and peroxidase. Peroxidase label can be visualized using DAB reaction product which becomes electron-dense after osmication. With the advent of colloidal gold particles as markers in immunocytochemical

I.B. Buchwalow and W. Böcker, *Immunohistochemistry: Basics and Methods*,
DOI 10.1007/978-3-642-04609-4_12, © Springer-Verlag Berlin Heidelberg 2010

studies (Faulk and Taylor 1971), ferritin and peroxidase labels became hardly ever used.

Colloidal gold is a suspension (or colloid) of sub-micrometre-sized particles of gold in a fluid, usually water. For electron microscopical immunolabeling, the gold particles are manufactured to any chosen size from 6 to 25 nm (see Sect. 2.4). Because of their high electron density, the gold particles are visible in the electron microscope without further treatment. Colloidal gold conjugates used for immuno-labeling at the electron microscope level consist of gold particles coated with a selected protein or macromolecule, such as an antibody, protein A or protein G (see Sect. 1.5). Because of their smaller size, protein A and protein G can have advantages over secondary antibodies. However, it should be noted that protein A and protein G do not recognize all classes of IgG, so a bridging secondary antibody of the corresponding IgG isotype may be required. Mouse monoclonal antibodies commonly have a stronger affinity to the chimeric protein A/G than to either protein A or protein G (see Table 1.1).

The size of a gold conjugate is co-responsible for labeling efficiency. The overall size is determined by the gold particle size as well as by the size of the proteins adsorbed onto the particle surface. When a specimen is relatively dense or intensely cross-linked, immunoreagents will be more hindered in their action. An improved labeling may result from using protein A (or G) or single Fab or F(ab)$_2$ fragment of the specific secondary antibody instead of the larger whole immunoglobulin mole-cule. A further improvement in labeling intensity can be achieved using smaller gold particles (1–5 nm), but the particles may not be so easily visualized in electron microscopes of limited magnifying power. With subsequent silver enhancing, gold particles become large enough to see even at much lower magnifications (see Sect. 2.4). The SPI-Mark Silver Enhancement kit (Structure Probe, Inc., http://www.2spi.com/) or AURION R-Gent system from Aurion (http://www.aurion.nl/) provide simple to use and sensitive systems for the amplification of immunogold labeled samples.

12.2   Fixation for Ultrastructural Immunohistochemistry

As for light microscopy, cell and tissue specimens must be fixed, but the specimen size must be less than 1 mm^3 and fixation should be kept short (30–120 min). Cell pellets can be previously embedded in agar gel (4%) to hold together for the following fixation and proceedings to resins. Traditionally, cell and tissue samples for electron microscopy are fixed in 2.5% glutaraldehyde. Due to the presence of two aldehyde moieties, glutaraldehyde introduces both intramolecular and intermo-lecular protein crosslinks, which results in excellent preservation of the ultrastruc-ture. However, steric hindrance resulting from extensive crosslinking of cellular proteins inhibits the antibody from reaching the intracellular epitopes. This fact leads to the choice of a compromise between good ultrastructural preservation and maintenance of the tissue antigenic properties. To achieve a satisfactory compromise between tissue preservation and accessibility of epitopes to antibodies, a mixture of formaldehyde (2–4%) and glutaraldehyde (0.05%–0.25%) may be recommended, although 4% neutral buffered formaldehyde with following post-fixation in OsO_4 also yields satisfactory results (Buchwalow et al. 2004, 2005; Groos et al. 2001; Stirling and Graff 1995). For biological tissue fixation, embedding and sectioning, we highly recommend the book by Glauert and Lewis, "Biological Specimen Preparation for Transmission Electron Microscopy" (Glauert and Lewis 1998).

Some authors recommend inactivating the residual aldehyde groups present after aldehyde fixation using incubation for 15 min with 0.05 M glycine or lysine in PBS buffer or with 0.1% $NaBH_4$ in PBS. We have not, however, found any experimental evidence that this step is really useful.

12.3 Resin Etching and Heat-Induced Antigen Unmasking in Resin Sections

Compared to epoxy (Epon or Araldite) tissue sections, acrylic (LR White or Low-icryl) tissue sections are less crosslinked and more hydrophilic than epoxy section, and their surface is rougher, which facilitates antigen detection without additional treatment. In contrast, epoxy sections require etching to unmask the antigen. Antigenic sites can be unmasked on epoxy thin sections by exposing the sections to strong oxidizing agents such as EDTA, hydrogen peroxide, sodium ethoxide or sodium metaperiodate. Sodium metaperiodate and sodium ethoxide appear to provide the most satisfactory results. This treatment enhances the hydrophilic nature of the embedding resin, allowing an easier antigen–antibody interaction, and additionally restores tissue antigenicity by removing osmium bonds from the sections of postosmicated tissues (Bendayan and Zollinger 1983). Thus, etching the sections is not only an oxidation/reduction step but also could act as an antigen-retrieval procedure. Subsequent additional heat-induced antigen retrieval (see Sect. 6.1.1) can enhance the labeling intensity to several fold (Stirling and Graff 1995). For more details of etching and antigen retrieval procedures, the readers are referred to Hayat (2002).

In our laboratory, we use sodium ethoxide for epoxy resin etching (Groos et al. 2001) with subsequent antigen retrieval in 10 mmol/l citric acid, pH 6.0, in a domestic pressure cooker (Buchwalow et al. 2004, 2005). The results of this procedure (see below) are shown in Figs. 12.1 and 12.2. Labeling can also be increased by heating in antigen-retrieval solution with a hotplate instead of using a pressure cooker (Stirling and Graff 1995). The intensity of the immunogold labeling can be increased almost three times when raising the temperature in the retrieval medium from 95°C to 135°C (Brorson 2001). A further increase in the immunolabeling of epoxy sections can be achieved by moderately increasing the proportion of the accelerator DMP-30 in Epon-embedding mixture (Brorson et al. 1999).

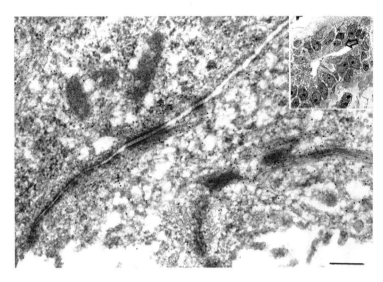

Fig. 12.1 Immunogold localization of glandular cytokeratins (Ck 8/18/19) in human mammary gland. The area labeled with *arrow* is represented at higher magnification. EPON embedding, etching with sodium ethoxide, antigen retrieval in citrate buffer, pH 6.0. 12 nm immunogold. Scale bar 0.20 μm

Fig. 12.2 Immunogold localization of endothelial NO synthase (NOS3) in skeletal muscles. **a** Immunogold labeling of NOS3 in subsarcolemmal mitochondria. **b** Inside of the sarcoplasm; NOS3 is localized along the contractile fibers, in the sarcoplasmic reticulum and in mitochondria. EPON embedding, etching with sodium ethoxide, antigen retrieval in citrate buffer, pH 6.0. 12 nm immunogold. 0.20 μm scale bar for entire layout

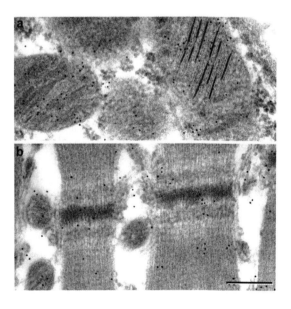

Sodium ethoxide* treatment for epoxy resin etching (Groos et al. 2001)
Hold grids for 10–20 s in an etching solution composed of saturated sodium ethoxide in absolute ethanol diluted to 50% with absolute ethanol.

Sodium ethoxide* treatment for epoxy resin etching (Stirling and Graff 1995)
Hold grids for 2 min in a fresh saturated solution of sodium ethoxide in absolute
ethanol (prepared overnight) diluted to 10% with absolute ethanol.
*Caution! Sodium ethoxide is a strong base. It reacts with water vigorously to give
ethanol, which is flammable, and sodium hydroxide, which is corrosive. Until
dissolved, keep in a tightly closed container, stored in a cool, dry, ventilated area.
Protect against physical damage and moisture. Isolate from any source of heat or
ignition. Separate from incompatibles, combustibles, organic or other readily
oxidizable materials. Wear protective gloves and use chemical safety goggles.

To get sodium ethoxide solution from sodium ethoxide powder and ethanol, mix it
slowly in a glass or stainless container and use all safety equipment (gloves,
eyewear, and ventilation). Always add powder to the liquid (!!!) and if it gets hot
to the touch, slow down. Both chemicals absorb water from the atmosphere, so
choose your location and the day appropriately (http://wiki.answers.com/).

Stirling and Graff (1995) preferred treatment with sodium metaperiodate followed by heating on citrate buffer. Sections were treated by heating them with a
hotplate at 95°C while they floated on 0.01 M citrate buffer (pH 6.0). Using this
combination, the authors reported a high probe density and sections remained
intact, with good ultrastructural detail.

Sodium metaperiodate** treatment for epoxy resin etching (Stirling and Graff
1995)
Incubate grids in a humid chamber on large drops of a saturated aqueous solution
of sodium metaperiodate (prepared overnight, 1 g in 5 ml Aqua dest) for 30–
60 min at room temperature.
**Warning! Do not ingest. Avoid contact with skin and clothing. Avoid breath-
ing dust. Store in tightly closed container. Avoid contact with combustible
materials. Use only with adequate ventilation. Wash thoroughly after handling.

In initial experiments, we found that the epoxy ultrathin sections were detached from
grids during etching in sodium ethoxide solution and by bubble formation when
boiled in a retrieval solution (10 mmol/l citric acid, pH 6.0). To overcome this
problem, we coated the grids dull side with Coat-Quick™ "G" Pen (Structure
Probe, Inc. http://www.2spi.com/) to improve the adhesion between tissue samples
and grids. We used nickel grids (200 Mesh Hexagonal, 3.05 mm, PLANO GmbH,
http://www.plano-em.de/) without coating with Formvar or Butvar. After heating, the
grids were allowed to cool by leaving them in the retrieval medium for 15 min. After
all treatments, grids were washed by dipping them repeatedly in water for 45–60 s.

Procedure for epoxy resin etching and antigen retrieval
– Apply epoxy ultrathin sections on nickel or gold grids coated with Coat-
  Quick™ "G" Pen (Structure Probe, Inc.) and correspondingly air dry.
– Rinse grids briefly in absolute ethanol.
– Hold grids for 10–20 s in an etching solution composed of saturated sodium
  ethoxide in absolute ethanol diluted to 50% with absolute ethanol.

- Rinse grids briefly in absolute ethanol, in 50% ethanol and then for 30 s or up to a few min in distilled water.
- Put grids in Eppendorf microtubes filled with an antigen retrieval solution (10 mmol/l citric acid, pH 6.0) and heat them in a domestic pressure cooker at boiling point for 2–5 min (Buchwalow et al. 2004, 2005). The cap of the microtube must have a hole bored by a needle in order to prevent the loss of the material in the process of bubble formation when boiled in a retrieval solution.
- Place the microtubes with grids at room temperature and allow them to cool for about 15–20 min. Before the beginning of the immunogold labeling protocol (see below), grids should be washed by dipping them repeatedly in water for 45–60 s.

12.4   Immunogold Labeling Procedure

Much literature continues to be published on the wide range of methods employed for tissue sections, resin-embedded or frozen (Beesley 1989; Polak and Van Noorden 1997; Renshaw 2007). The protocol provided in this section is based on guidelines for the indirect (two-step) immunogold labeling with BioSite Colloidal Gold Reagents (Energy Beam Sciences, Inc., http://www.ebsciences.com/company.htm). This protocol may also be used as a guide to labeling with colloidal gold conjugates available from other vendors such as Structure Probe, Inc. (http://www.2spi.com/), Aurion (http://www.aurion.nl/), Electron Microscopy Sciences (http://www.emsdiasum.com/microscopy/default.aspx), Polysciences, Inc. (http://www.polysciences.com), Dianova (http://www.dianova.de) and others.

Immunogold labeling procedure
- Tissue processing: tissue specimen (0.5–1.0 mm^3) are fixed in 4% buffered formalin for 30–60 min, post-fixed in 1% osmium tetroxide in cacodylate or phosphate buffer, pH 7.2–7.4, stained en bloc for 30 min with 2% aqueous uranyl acetate*, then dehydrated in ethanol and embedded in epoxy or acrylic resin.
- Sections are best mounted on nickel or gold grids** of high mesh number to improve adherence during incubations. Plastic film coating of the grids (e.g. Formvar or Butvar) is optional and depends on the stability of the sections.
- Carry on resin etching and antigen-retrieval steps (see above). Useful for epoxy sections but not for acrylic (LR Wite) sections.
- Grids are floated on drops of various buffer and antibody solutions usually on strips of dental wax or Parafilm in a Petri dish. All steps may be performed at room temperature. Do not allow the sections to dry out during this procedure since this may give rise to non-specific charge attachment of the antibodies.
- Place the grid, section face down, on a 25 µl droplet of PBS or TBS containing 5% normal serum of species in which the secondary antibodies were raised and incubate for 10 min***.

- Use a platinum wire loop or forceps to transfer the grid to the surface of a droplet of appropriately diluted primary antibody****. Incubate from 30 min to overnight (depending on dilution, temperature, etc.). Longer incubations with higher dilutions of antibody produce more specific labeling.
- Transfer the grid to a series of droplets of PBS (5 × 2 min).
- Transfer the grid to a droplet of gold conjugate diluted 1:10 to 1:100 (or greater) in PBS or Tris-buffered saline. Incubate for 1 h.
- Transfer the grid to a series of droplets of distilled water (5 × 2 min) to wash away unbound gold conjugates.
- Stain embedded sections lightly in uranyl acetate and lead citrate (optional). Frozen sections may be stained with osmium tetroxide vapor. Staining with OsO_4 or uranyl acetate may be conducted without obscuring the gold particles. Wash and examine under the electron microscope.

Note: *Alternatively, uranyl acetate staining may be carried out after immunolabeling.

**Nickel or gold grids are preferred to copper ones for immuno incubations, since copper will be corroded during epoxy resin etching and antigen retrieval steps. Additionally, nickel and gold are more inert and less poisonous to immunoreactions. Nickel grids can, however, be annoying because of their magnetic properties. This is easily overcome by using either non-magnetic tweezers or by using a flattened loop to transfer grids from droplet to droplet during immunoincubations.

***Ready-to-use blocking solutions containing normal inactivated sera or BSA for use with colloidal gold conjugates are available from Aurion (http://www. aurion.nl/). When labeling with protein A and protein G conjugates, this step can be omitted.

****When labeling epoxy resin sections that were subjected to etching and antigen retrieval, antibody dilutions are usually about the same as for light microscopy. When labeling tissue sections embedded in acrylic resins such as LR White or Lowicryl, primary antibody should be about ten times more concentrated than working dilution for light microscopy.

5% Uranyl Acetate Solution
To prepare 50 ml, add 2.5 g of uranyl acetate to 50 ml of distilled water. Cover with foil and stir overnight. Add 10 drops of glacial acetic acid. Store solution at 4°C.

Reynold's Lead Citrate Solution
To prepare 50 ml, add chemicals in distilled water in following order

Lead nitrate	−1.33 g
Sodium citrate, dihydrate	−1.76 g (solution becomes cloudy when sodium citrate is added)
1 N NaOH	−5 ml (solution becomes clear when NaOH is added)
Distilled water	−30 ml

Stir for 10 min to dissolve and add additional 15 ml of distilled water. Store solution for 3–6 months at 4°C. Note: the amount of NaOH is very important. Too much will make solution cloudy.

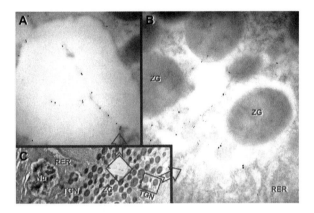

Fig. 12.3 Co-localization of kinesin (6 nm gold) and tubulin (15 nm gold) in the lumen and trans-Golgi network in the rat pancreas acinus. An overview of the acinus is presented in the inset (**c**) at lower magnification. Boxes in the inset (**c**) denote regions of lumen (**a**) and trans-Golgi network (**b**) at higher magnification. Abbreviations: *TGN* trans-Golgi network, *Nu* nucleus, *ZG* zymogen granules, *RER* rough endoplasmic reticulum

This protocol is also applicable for double or multiple immunolabeling using secondary antibodies labeled with gold particles of different size. To facilitate visual discrimination, the overlap between two different sizes of gold particles must be at least twofold (e.g., 6 and 15 nm or 10 and 25 nm). By double labeling, use the smaller gold conjugate for detecting the antigen with a lower expression level. Primary antibodies must be of different species or of different IgG isotypes (see Chap. 8). According to this protocol, primary antibodies of different species and IgG isotypes can be used in a mixture and not necessarily applied subsequently.

Figure 12.3 demonstrates colocalization of kinesin and tubulin in the rat pancreas acinus. Pancreatic tissue specimen (0.5–1.0 mm^3) were fixed in 4% formaldehyde in PBS overnight at $+4°C$, washed three times in PBS for 15 min, subsequently dehydrated in ethanol and embedded in LR White. Ultrathin sections (50–100 nm) were incubated overnight at $+4°C$ with anti-kinesin mouse monoclonal antibodies and with rabbit polyclonal antibodies to tubulin. Bound primary antibodies were labeled for 2 h at room temperature with anti-mouse IgG and anti-rabbit IgG goat secondary antibodies conjugated with colloidal gold particles (6 and 12 nm, respectively), washed three times for 10 min in PBS and distilled water, counterstained with uranyl acetate and finally examined with a CM 10 transmission electron microscope (Philips, Hamburg, Germany).

Opening this section, we wrote that immunohistochemistry at the ultrastructural level is based on the same principles as immunohistochemistry for light microscopy. The same is also true for the specificity controls. Obligatory controls are (1) omission of incubation with primary antibodies, (2) substitution of primary antibodies by the corresponding IgG at the same final concentration, and (3) incubation in media containing primary antibodies that have been preincubated at room temperature for 2 h with a ten-fold molar excess of the corresponding control peptide used as an immunogen (so-called pre-absorption control). And last but not least, one should never start immunohistochemistry at the ultrastructural level without first attesting the primary antibody under the light microscope.

References

Beesley J (1989) Colloidal gold: a new perspective for cytochemical marking. Royal Microscopical Society Handbook No 17. Oxford Science Publications, Oxford

Bendayan M, Zollinger M (1983) Ultrastructural localization of antigenic sites on osmium-fixed tissues applying the protein a-gold technique. J Histochem Cytochem 31:101–109

Brorson SH (2001) Heat-induced antigen retrieval of epoxy sections for electron microscopy. Histol Histopathol 16:923–930

Brorson SH, Andersen T, Haug S, Kristiansen I, Risstubben A, Tchou H, Ulstein J (1999) Antigen retrieval on epoxy sections based on tissue infiltration with a moderately increased amount of accelerator to detect immune complex deposits in glomerular tissue. Histol Histopath 14:151–155

Buchwalow IB, Podzuweit T, Samoilova VW, Wellner M, Haller H, Grote S, Aleth S, Boecker W, Schmitz W, Neumann J (2004) An in situ evidence for autocrine function of NO in the vasculature. Nitric Oxide 10:203–212

Buchwalow IB, Minin EA, Samoilova VE, Böcker W, Wellner M, Schmitz W, Neumann J, Punkt K (2005) Compartmentalization of NO signaling cascade in skeletal muscles. Biochem Biophys Res Comm 330:615–621

Faulk WP, Taylor GM (1971) Immunocolloid method for electron microscope. Immunochemistry 8:1081–1092

Glauert AM, Lewis PR (1998) Biological Specimen Preparation for Transmission Electron Microscopy. In: Glauert AM (ed). Practical Methods in Electron Microscopy, vol 17. Portland Press, London

Groos S, Reale E, Luciano L (2001) Re-evaluation of epoxy resin sections for light and electron microscopic immunostaining. J Histochem Cytochem 49:397–406

Hyatt MA (2002) Microscopy, immunohistochemistry, and antigen retrieval methods: for light and electron microscopy. Plenum Publishers, New York

Polak JM, Van Noorden S (1997) Introduction to Immunocytochemistry. BIOS Scientific, Oxford

Renshaw S (2007) Immunohistochemistry. Scion Publishing, Cambridge

Saga K (2005) Application of cryofixation and cryoultramicrotomy for biological electron microscopy. Med Mol Morphol 38:155–160

Stirling JW, Graff PS (1995) Antigen unmasking for immunoelectron microscopy: labeling is improved by treating with sodium ethoxide or sodium metaperiodate, then heating on retrieval medium. J Histochem Cytochem 43:115–123

Useful Websites

http://www.nanoprobes.com/TechIEM.html
http://www.plano-em.de/
http://www.ebsciences.com/company.htm
http://www.2spi.com/
http://www.aurion.nl/
http://www.emsdiasum.com/microscopy/default.aspx

Chapter 13
Diagnostic Immunohistochemistry

Routine hematoxylin and eosin-stained sections are often appropriate and adequate for the definite diagnosis of pathological lesions. However, in cases where light microscopic examination of tissue sections from biopsy and surgical specimens is inconclusive, immunohistochemistry can efficiently back up histopathology. Albert H. Coons was the first who used a specific antibody to localize pneumococcal antigen in tissues (Coons et al. 1941, 1942). Since that time, the use of antibodies to detect individual antigens in situ has developed into a powerful diagnostic tool. With the broad availability of immunological markers, immunohistochemistry is becoming increasingly indispensable in pathology and in diagnostic pathology of different organ systems. At present, most of the antigens are easily detected both in formalin-fixed paraffin-embedded tissue samples and in ethanol-fixed cytological smears. Over the past 20 years, immunohistochemical and immunocytological stains have therefore become an integral part of routine diagnostic procedures in difficult cases.

Currently in about two-thirds of these difficult cases, an appropriate diagnosis can be made due to immunohistochemical stains with the use of antibodies. Antibodies for diagnostic use can be grouped in the following categories or areas: (1) antibodies for tumor typing, for lymphocytic typing of subgroups, (2) hormones and hormone receptors, peptides, amines, (3) oncofetal antigens, (4) oncogenes and their products, (5) cell proliferation antigens, (6) infectious agents, and (7) immunoglobulins and complement fragments for the study of glomerular diseases. For a wide-ranging review of this subject, the readers are referred to a comprehensive textbook "Diagnostic Immunohistochemistry" (Ed. by Dabbs 2006) specifically designed to target diagnostic dilemmas in surgical pathology of different organ systems. In frames of this introductory handbook, we restrict our discussion of this subject to applications of triple immunofluorescence stainings for keratin subtypes and some myoepithelial markers in some areas of proliferative breast disease and some special tumors of breast.

I.B. Buchwalow and W. Böcker, *Immunohistochemistry: Basics and Methods*, 109
DOI 10.1007/978-3-642-04609-4_13, © Springer-Verlag Berlin Heidelberg 2010

13.1 Keratins and Myoepithelial Markers as Diagnostic Markers in Proliferative Breast Disease and in Tumors of Breast

Keratins are alpha-type fibrous polypeptides with a diameter of 7–11 nm. They are important components of the cytoskeleton in almost all epithelial cells as well as in some nonepithelial cell types. Keratins are generally held to be the most ubiquitous markers of epithelial differentiation. So far, 20 distinct types numbered by Moll et al. (1982a, 1990, 1992) have been revealed. Keratins were earlier thought to be separable into "hard" and "soft," or "cytokeratins" and "other keratins," but these designations are now understood to be incorrect. In 2006, a new nomenclature (Schweizer et al. 2006) was adopted for describing keratins which takes this into account (Table 13.1).

This section summarizes the current information about different human keratins, their functional significance, cell-type-specific distribution as well as their coexpression with myoepithelial markers and estrogen receptor alpha (ER), because these markers exhibit characteristic expression patterns in some human breast tumors. In many of these cases, a definite diagnosis can be achieved only when additional information is provided by immunohistochemistry.

13.1.1 Human Keratins and Their Expression Patterns in Breast Epithelium

The new consensus nomenclature (Table 13.1) comprises type I keratins K9–K10, K12–K28, and K31–K40 (including K33a and K33b) and type II keratins K1–K8 (including K6a, K6b and K6c) and K71–K86 (Moll et al. 2008). Keratins show a characteristic pairing of heterodimers between one type I keratin and one type II keratin. In the non-stratified (simple) epithelia, "simple–epithelial keratins" such as K7/K8/K18 form sparse and loosely distributed keratin filaments, while in contrast squamous epithelia and especially cornified stratified epithelia form keratinocyte-type keratins such as K5/K14 in the basal layer and – with even more pronounced bundling – K1/K10 in the suprabasal layers and K2/K10 in the uppermost ones densely bundled as tonofilaments. This clearly underscores the importance of the keratins for the tissue integrity and the relevance of the molecular diversity of keratin proteins. They are also characteristic of distinct – including the terminal – stages during cellular epithelial differentiation from embryonal to adult or of the internal maturation program during development.

13.1.2 Progenitor Cell Keratins K5/K14

The type II keratin K5 and the type I keratin K14 form the primary keratin pair of progenitor cells of the breast epithelium, including cells of the basal

Table 13.1 The new human keratin nomenclature (Schweizer et al. 2006; Moll et al. 2008)

Keratin types	Type I		Type II	
	New name	Former name	New name	Former name
Epithelial keratins	K9	K9	K1	K1
	K10	K10	K2	K2
	K12	K12	K3	K3
	K13	K13	K4	K4
	K14	K14	K5	K5
	K15	K15	K6a	K6a
	K16	K16	K6b	K6b
	K17	K17	K6c	K6e/h
	K18	K18	K7	K7
	K19	K19	K8	K8
	K20	K20	K76	K2p
	K23[a]	K23	K77	K1b
	K24[a]	K24	K78[a]	K5b
			K79[a]	K61
			K80[a]	Kb20
Hair follicle-specific	K25	K25irs1	K71	K6irs1
epithelial keratins (root	K26	K25irs2	K72	K6irs2
sheath)	K27	K25irs3	K73	K6irs3
	K28	K25irs4	K74	K6irs4
			K75	K6hf
Hair keratins	K31	Ha1	K81	Hb1
	K32	Ha2	K82	Hb2
	K33a	Ha3-I	K83	Hb3
	K33b	Ha3-II	K84	Hb4
	K34	Ha4	K85	Hb5
	K35	Ha5	K86	Hb6
	K36	Ha6		
	K37	Ha7		
	K38	Ha8		
	K39	Ka35		
	K40	Ka36		

[a]Expression pattern still unknown, only gene information available

(Purkis et al. 1990) and the luminal layer (Böcker et al. 2002; Böcker et al. 2006). Triple immunofluorescence studies have clearly shown that K5/14+progenitor cells differentiate to glandular (k8/18+) and myoepithelial (sma+. Others+) cells (Böcker et al. 2002; Böcker et al. 2006) through intermediate glandular (K5/14+; K8/18+) and myoepithelial (K5/14+;sma+) cells (Figs. 13.1–13.4). This is one of the classical examples of the carefully regulated differentiation-specific expression of keratin proteins. In the literature basal keratins are often interpreted as myoepithelial markers *strictu sensu*. This is flat not true. The studies cited above clearly show that the basal-type keratins are per se not myoepithelial markers, because they also stain noncommitted progenitor cells (K5/14 only), glandular progenitor (K5/14 only) and intermediate glandular cells (K5/14; K8/18).

A similar process of differentiation has been demonstrated in the basal cell layer of the squamous epithelium of the skin which seems to contain the stem cells

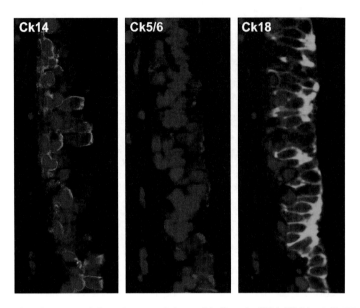

Fig. 13.1 Triple immunostaining of a normal duct epithelium for K14 (Ck14, *pink*), K5 (Ck5/6, *red*), and K8/18 8 (Ck18, *green*). Note that few progenitor cells are stained only for K5 and/or K14 (1), whereas most are stained both for basal (usually K5) and glandular keratins K8/18. These cells represent intermediary glandular cells. Few cells stained only for K8/18 are differentiated glandular cells

(Fuchs and Green 1980) and which differentiate to squamous cells in suprabasal cell layers with a shift of K5/14+ to K1/10+ cells (Fig. 13.5). In basal cells of the prostate (Moll et al. 1982a) and in the ducts of the salivary gland K5/14+ cells are also found, which we believe are progenitors of both glandular and myoepithelial cells in this organ. We further hypothesize that in all these tissues K5/14+ cells form the cell compartment which is closely linked to the stem cells in these organs.

Several monoclonal antibodies (MAbs) specific for K5 and K14 have been described that help to reveal their exact tissue distribution. These antibodies include MAb AE14 (Moll et al. 1989) against K5, and MAbs LL001 and LL002 against K14 (Purkis et al. 1990). The best performance for K5 on paraffin sections is displayed by MAb D5/16B4 (Lobeck et al. 1989; Demirkesen et al. 1995) which is however often regarded as "K5/K6 antibody" (Böcker et al. 2006 and citations therein).

13.1.3 K17: Keratin of Basal/Myoepithelial Cells

In normal breast epithelium the type I keratin K17, which was identified by Moll et al. (1982b) by early gel electrophoretic studies as a major keratin of basal cell carcinomas of the skin, shows a similar distribution in breast epithelium as K5

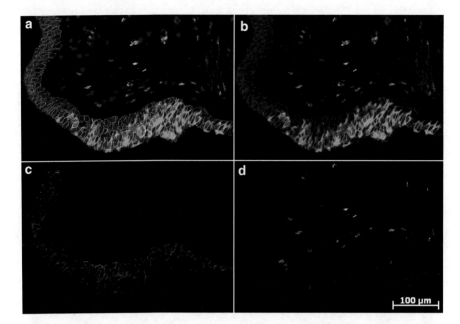

Fig. 13.2 Triple staining of a nipple duct and the adjacent nipple epidermis for K14 (*red*), ER (*pink*), and K8/18 8 (*green*). **a** Composite image. Nuclei are counterstained with DAPI (*blue*). **b** The same as a **a**, but without nuclear counterstaining. **c** Red component (for K14) of the composite image. **d** Pink component (for ER) of the composite image. Note the immature glandular epithelium of the duct and the abrupt transformation to squamous cell epithelium of the nipple. ER (*pink* nuclei) is only expressed in glandularly differentiated cells

(Moll et al. 1982a, 1983; Troyanovsky et al. 1989, 1992). Thus, K17 may also be regarded as a "basal/myoepithelial cell keratin" which probably is mainly expressed in progenitor cells.

13.1.4 Glandular Keratins

K8/K18: primary keratins in luminal (glandular) epithelium: Keratins K8 and K18 are typically coexpressed and constitute the primary keratin pair in the breast epithelium (Franke et al. 1981; Moll et al. 1982b; Owens and Lane 2003; Böcker et al. 2002; Böcker et al. 2006 and citations therein). Thus we regard these keratins as surrogate for a glandular differentiation. The glandular cells (K8/18+only) probably derive from glandularly committed K5/14+progenitor cells through intermediate stages (K4/14+; K8/18+) (Figs. 13.1–13.3). The glandular lineage can further be subdivided into ER alpha+ and ER alpha− cells.

K7/K19: secondary keratins: Apart from K8/K18, type II keratin K7 and type I K19 are "additional" (secondary) and also widely distributed glandular keratins

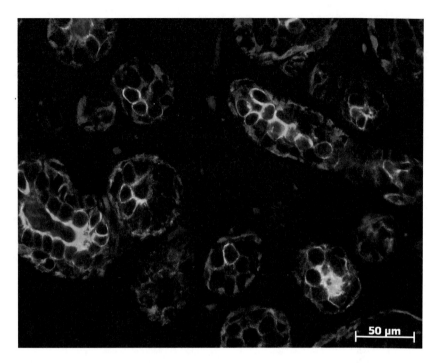

Fig. 13.3 Triple immunostaining of a an immature lobule for K5 (*red*) and K8/18 (*green*), and SMA (*pink*); acini contain mainly immature epithelium (K5, *red*) and only few differentiated cells express K8/18 (*green*)

which are coexpressed. They typically occur as a keratin pair. The type I keratin K19 is the smallest keratin and is exceptional since it widely lacks the nonalphahelical tail domain typical of all other keratins.

13.1.5 K1/K10: Major Keratins of Keratinocyte Differentiation and Keratinization

Physiologically, these keratins are found in the nipple epidermis. In the nipple, the typical immature multilayered epithelium of large ducts ends abruptly, to give way to the typical epidermis in the process of terminal differentiation and keratinization, which is characterized by a profound change in keratin expression (Fig. 13.5). This involves a switch from expression of the basal cell keratins (K5, K14) to the suprabasal epidermal keratins, the type II keratin K1 and subsequently the type I keratin K10. Ultrastructurally, keratin filaments composed of the pair K1/K10 form particularly dense bundles which are so characteristic of suprabasal epidermal keratinocytes (Fuchs and Green 1980; Moll et al. 1982a). Among various antibodies against K1 and K10 described in the literature, MAb DE-K10

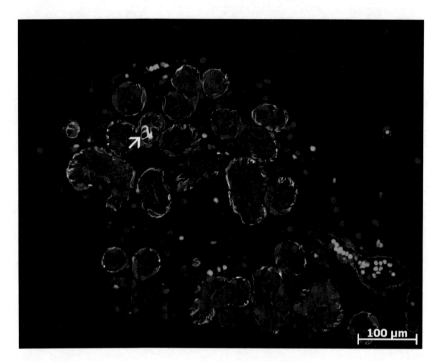

Fig. 13.4 Triple immunostaining of a mature lobule for K8/18 (*red*), K5 (*green*) and SMA (*pink*). Note that most of the acini contain K8/18; only one acinus contains intermediate cells (*arrow, hybrid orange color*)

against K10 is particularly suitable for application with paraffin-embedded tissues (Ivanyi et al. 1989).

13.1.6 Myoepithelial Differentiation Markers

Myoepithelial cells are contractile elements found in salivary, sweat, and mammary glands that show a combined smooth muscle and epithelial phenotype (Foschini et al. 2000). In the normal breast, the ductal and acinar units are lined by two cell layers: the inner layer lining the lumen and an outer layer of myoepithelial cells. An intact myoepithelial cells layer is seen in both benign and in situ lesions, whereas loss of the myoepithelial cells layer is considered the rule for the diagnosis of invasive cancer (Kalof et al. 2004 and citations therein).

Because myoepithelial cells are not always readily identifiable on routine hematoxylin- and eosin-stained sections, many immunohistochemical methods have been used to highlight an intact myoepithelial cells layer. Recent studies have reported CD10 and smooth muscle myosin heavy chain (SMMHC) expression in

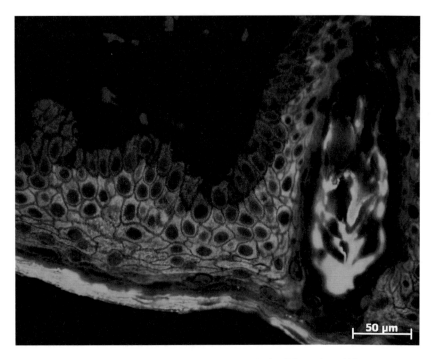

Fig. 13.5 Double immunostaining of the nipple epidermis for K5 (*pink*) and K10 (*green*). Note that with increasing differentiation, K5 is lost and the cells express K10

myoepithelial cells of the breast, supporting their use as markers to help distinguish invasive breast carcinoma from ductal carcinoma in situ. CD10 is a 100 kD cell surface metalloendopeptidase called neprilysin which inactivates a variety of biologically active peptides. It was initially identified as the common acute lymphoblastic leukemia antigen. Subsequent studies, however, have shown that CD10 is expressed on the surface of a wide variety of nonlymphoid cells and other tissues, such as breast myoepithelial cells. The antibody to SMMHC is also reactive for smooth muscle cells and myoepithelial cells. Being a structural component of the smooth muscle myosin molecule, SMMHC is used as a specific marker of "terminal" smooth muscle differentiation. Thus the immunohistochemical stainings of these two proteins have to be considered in the cellular context of a given tissue or lesion. In the basal layer of the normal breast epithelium and in proliferative breast lesions they indicate a myoepithelial differentiation and are often co-expressed with basal keratins (Fig. 13.1). A recent comparative study of CD10 and SMA expression in myoepithelial cells of the breast concluded that CD10 was uniformly positive in myoepithelial cells of normal breast and may serve as a useful marker of breast myoepithelial cells in difficult breast lesions (for example, sclerosing adenosis versus tubular carcinoma) (Moritani et al. 2002).

Protein p63 is a member of the p53 protein family. This protein is not a specific myoepithelial marker as described in the preceding paragraph, but it is often expressed in the nuclei of myoepithelial cells positively immunostained for basal keratins. Nuclear localization of p63 was found well correlated with nuclear accumulation of p53 (Narahashi et al. 2006). p53 is a tumor suppressor gene that, when working normally, helps to stop cells becoming cancerous. Increased p53 expression is a frequent finding in malignant tumors. The predominant localization of p63 is in the basal layer of stratified squamous and transitional epithelia (Yang et al. 1998). These basal cells act as the progenitors of the suprabasal cells, which undergo differentiation and cell death in regenerative epithelia (Jetten and Harvat 1997). High levels of p63 are found in basal and proliferating cells of many tissues, especially in stem cells. The high expression levels of p63 in "basal cell like" breast cancers suggest that they may originate from basal or myoepithelial cells (Assefnia et al. 2006). In normal conditions, p63 is expressed in basal cells of the prostate (reserve cells), in the salivary gland (both in reserve cells of striated and excretory ducts), in myoepithelial cells of intercalated duct and acini, and in basal and luminal cells of the breast epithelium.

13.1.7 Keratins and Myoepithelial Markers as Diagnostic Markers in Proliferative Breast Disease and in Tumors of Breast

One of the important fields of "application" of keratins (making use of their special and characteristic expression patterns in distinct cell types, differentiation stages or functional states) is their use – aided by specific myoepithelial antibodies – as immunohistochemical markers in diagnostic tumor pathology. Epithelial tumors maintain – at least widely – specific features of the keratin expression patterns of the respective cell type of origin, thus helping to identify and classify a given tumor. A relatively small panel of keratin antibodies has attained diagnostic importance in breast pathology with the use of selected antibodies or as part of a panel together with myoepithelial and functional markers. This approach has become a diagnostic standard in state-of-the-art clinical pathology (see recent overviews Chu and Weiss 2002b; Dabbs 2006).

13.1.8 Carcinomas of Luminal Phenotype (Luminal Type Carcinomas) in the Breast

As regards malignant tumors, K8 and K18 (and also K7 and K19) are expressed in most breast carcinomas (breast carcinomas of luminal phenotype; luminal-type breast carcinomas) as the only keratins. Most of these tumors express estrogen receptor alpha and comprise about 90% of all cases of breast cancer; most breast carcinomas therefore constitutively express the classical glandular keratins K7, K8,

K18 and K19 (Altmannsberger et al. 1986; Malzahn et al. 1998) including ductal and lobular in situ neoplasia (Böcker et al. 2006 and citations therein). This approach permits to differentiate the following invasive luminal-type carcinomas of the breast:

- Invasive ductal (NST)
- Invasive lobular
- Tubular
- Cribriform
- Mucinous
- Papillary
- Apocrine
- others

13.1.9   Tumors of Basal Phenotype

Microarray-based expression profiling of breast carcinomas has led to the definition of distinct subgroups. One of these has been designated as basal-like group (Sørlie et al. 2001), which is characterized by relatively poor prognosis. These tumors are usually grade 3 and, often but not always, triple negative (ER-, PR-, and Her2-negativity) (Pia-Foschini et al. 2003; Gusterson et al. 2005; Jacquemier et al. 2005; Reis-Filho and Tutt 2008). It was suggested to define the basal-type cancers as those characterized by the expression of basal cell-typical keratins K5, K14 and/or K17 (Abd El-Rehim et al. 2004). We fully support this suggestion, because immuno-histochemistry is currently the most distributed tool available for pathologists and the most robust and standardized methods available to evaluate the basal character. Several older reports have already described basal-type carcinomas (Moll et al. 1983; Wetzels et al. 1991), and some already mentioned the bad prognosis (Dairkee et al. 1987; Malzahn et al. 1998). This has now been confirmed and extended by the growing body of recent microarray data, making it possible to define a subgroup of breast cancers with a bad prognosis and association with BRCA1 mutation or dysfunction (Gusterson et al. 2005; Diaz et al. 2007). Thus immunostaining for keratins such as K5 may become important (van de Rijn et al. 2002; for further references, see Gusterson et al. 2005).

In addition, most salivary gland like tumors of the breast are basal-type (k5/14+progenitor cell derived) tumors with glandular or myoepithelial differentiation or with a heterologeous squamous or "mesenchymal" differentiation (Böcker, unpublished data). Basal-type carcinomas in the breast and its corresponding counterparts in the salivary gland (in brackets below) are represented by the following subtypes:

- Pleomorphic adenoma (same)
- Basal cell adenoma (same)
- Syringoma (not known in parotid gland)
- Adeno-myoepithelial tumors (epithelial-myoepithelial tumors)
- Myoepithelial tumors (same)

- Adenoid-cystic carcinoma (same)
- Adenosquamous carcinoma (same; mucoepidermoid)
- Squamous cell carcinoma (same)

In all salivary gland-like tumors of the breast (notably basal adenomas, pleomorphic adenoma, adeno-myoepithelial tumors (AMT), myoepithelioma, adenoid-cystic carcinoma (ACC), adeno-squamous carcinoma (ASC), mucoepidermoid carcinoma of the salivary gland, squamous cell carcinomas), we include here syringomatous adenoma and metaplastic carcinoma of the breast, cells with K5/K14 and/or K17 expression are the constituting cell types, from which glandular and myoepithelial differentiation takes evolves. We could also show that squamous cell and mesenchymal cell differentiation in metaplastic carcinoma also includes a transformation of K5/14-progenitor cell to the metaplastic squamous or mesenchymal cell lineage. Glandular keratins 7/8/18/19 and/or squamous cell keratins or even the intermediate filament vimentin, which is characteristic of mesenchymal differentiation, may be conditionally expressed depending on the type of tumor. Highly sensitive monoclonal antibodies against all these antigens are available, such as the classical mouse monoclonal CAM5.2 clone against K8 and clone Ks18.04 against K18 (Bártek et al. 1991), smooth muscle actin (SMA), SMMHC and vimentin. These antigens may be helpful in diagnostic immunohistochemistry in cases of such lesions, to prove their nature. Figures 13.6–13.8 demonstrate a typical case of adenoid-cystic carcinoma.

Furthermore, much interest has evolved regarding the role of K5/K14 in breast pathology in several aspects, especially in the classification of benign proliferative lesions and ductal and obular in situ neoplasia (Otterbach et al. 2000). A typical example of a benign epithelial proliferation is shown in Fig. 13.7.

13.1.10 Squamous Cell Carcinomas

The expression spectrum of K5 and K14 of squamous cell carcinoma of the breast corresponds well to the patterns of squamous cell carcinomas in other parts of the

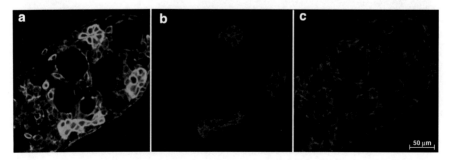

Fig. 13.6 Triple staining of an adenoid-cystic carcinoma for K14 (*green*: **a**), (**b**) for K8/18 (*red*: **b**) and for SMA (*pink*: **c**). Note that the k14+ cells differentiate to glandular and myoepithelial cells

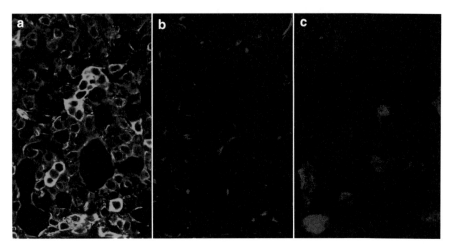

Fig. 13.7 Triple immunostaining of an adenoid-cystic carcinoma for K14 (*green*: **a**), for SMA (*pink*: **b**), and for collagen IV (*red*: **c**). The collagen IV is surrounded by myoepithelial cells (*pink*) which probably synthesize and secrete this protein

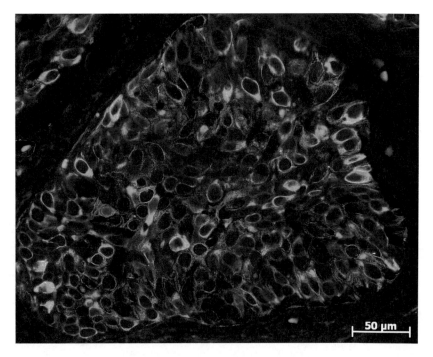

Fig. 13.8 Triple staining of a usual ductal hyperplasia for K5 (*red*), K8/18 (*green*) and SMA (*pink*). Note that the proliferating cells are only glandular cells. Myoepithelial cells are seen only in the outer layer (*pink*) and do not participate in the proliferation process

body and furthermore in normal squamous epithelium. In this view, it seems reasonable to regard K5/14+progenitor cells as the cells or origin of both squamous cell metaplasia (Fig. 13.9) and squamous cell carcinoma in breast parenchyma. Such tumors strongly express these keratins, whereas little, focal, or no expression is found in luminal-type carcinomas of the breast (Moll et al. 1982a, 1989; Moll 1998; Chu and Weiss 2002a, b). Focal K5 expression may be observed in certain adenocarcinoma types, notably in adenocarcinomas of the endometrium, the ovary, and the pancreas, which seems to be related to their potency for focal squamous differentiation (Moll 1998; Chu and Weiss 2002a, b).

In squamous cell carcinomas, focal expression of K1 and K10, usually in relation to maturation and keratinization, can be observed regardless of whether they are derived from the skin or from internal organs (for references, see Moll 1998). However, quantitatively, squamous cell carcinomas embark on an alternative maturation pathway characterized by abundant expression of K6 and K16. In summary, basal-type keratins K5 and K6 are useful as general markers for tumors derived from basal (K5/14+ progenitor) cells, whereas K1 and K10 can be regarded as "keratinization markers" and therefore as squamous differentiation markers.

Squamous cell carcinomas of different sites of origin are generally characterized by a predominance of stratified-epithelial/keratinocyte-type keratins but may coexpress certain simple-epithelial keratins (for details, see Moll 1998). Most of these

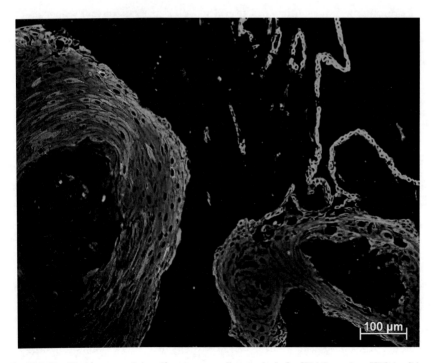

Fig. 13.9 Double immunostaining of squamous cell metaplasia for K14 (*green*) an K10 (*red*) in a papilloma of the breast

tumors strongly express keratins K5, K14 and K17, normally found in the basal layer, as well as keratins K6 and K16 characteristic of hyperproliferative keratinocytes. Focally, there may be expression of K1/K10 (particularly in more differentiated tumor cells). The coexpression of simple epithelia-typical keratins comprises K8, K18, and K19, and different studies have suggested that this coexpression seems to be more pronounced in poorly differentiated squamous cell carcinomas (for references, see Moll 1998). Recently it has been demonstrated that the expression of K8 and K18 in squamous cell carcinomas of the oral cavity is an independent prognostic marker and indicates a decreased overall and progression-free survival (Fillies et al. 2006).

As our knowledge of the molecular basis of proliferative breast diseases and tumors has increased, immunohistochemistry is being used with increasing frequency to identify underlying molecular changes or the presence of specific molecular markers in tumors, both as an aid to diagnosis and as a guide to appropriate therapy. In many cases of undifferentiated tumors, a definite diagnosis can be achieved only when additional information is provided by immunohistochemistry.

13.2   Tissue Microarrays

Tissue microarrays are particularly useful in quick analysis of many cancer samples. Conventional histological techniques allow sections from only one probe per slide to be stained or immunolabeled on each occasion. The analysis of a large number of tumor tissues with conventional techniques of molecular pathology is tedious and slow. Local variations in tissue fixation, antigen retrieval, and staining methods may significantly affect the staining profile obtained with a given antibody. To facilitate the proper diagnostic procedures and quality assurance in diagnostic immunohistochemistry, a simple histological technique was developed that increases the number of probes on each slide. The method that meets these requirements was first described by Battifora in 1986 (Battifora 1986) and later in by Kraaz et al. (1988). Using this method representative areas of tumors or tissues are delineated on routinely stained sections retrieved from the surgical pathology files. Corresponding areas in the paraffin wax blocks are then punched with a modified skin biopsy instrument (4 mm in diameter). Multiblocks containing up to 30 punch specimens from different tumors or tissues are made by placing the punched specimens in a warm cast containing a small amount of melted paraffin wax. The position of each punch specimen in the multiblock is recorded. Each multiblock section is used for one antibody. Only tumors showing various degrees of positivity and negative reactive tissues are chosen for this purpose. Sections of the specimen to be immunostained are then mounted on the same multiblock section slide containing the necessary controls. Five to six punch control specimens are often sufficient in each multiblock, which thus reduces the amount of antibody necessary for coverage. The economical aspect of examination is also important (Hsu et al. 2002).

In 1998 Kononen and collaborators (Kononen et al. 1998) further developed this technique, which uses a novel sampling approach to produce tissues of regular size and shape that can be more densely and precisely arrayed. Since that time this method has acquired a new name, tissue microarrays (TMAs), and represents a significant advance over earlier multitissue blocks in terms of the number of samples and uniformity of sample size.

In this technique (http://en.wikipedia.org/wiki/Tissue_microarray), a hollow needle is used to remove tissue cores of various diameter (from as small as 0.6–4 mm) from regions of interest in paraffin-embedded tissues (clinical biopsies or tumor samples) (Figs. 13.10, 13.11).

These tissue cores are then inserted in a recipient paraffin block (Fig. 13.12) in a precisely spaced array pattern. Each microarray block can be cut into from 100 to 1000 sections (Fig. 13.13), which can be subjected to independent tests with immunolabeling (Moch et al. 2001).

The tissue microarray technology has the potential to significantly accelerate molecular studies that seek associations between molecular changes and clinico-pathologic features of the cancer. Hundreds or thousands of tissue cores can be arranged on a single slide, and then analyzed by a single immunostaining or in situ hybridization reaction. Sections of the microarray also provide targets for parallel in situ detection of DNA, RNA and protein targets in each specimen on the array, and consecutive sections allow the rapid analysis of hundreds of molecular markers in the same set of specimens. With detection of six gene amplifications as well as p53 and estrogen receptor expression in breast cancer, the group of Juha Kononen

Fig. 13.10 A tissue microarray instrument with a paraffin wax block

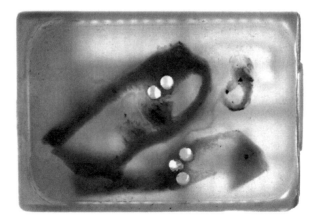

Fig. 13.11 The paraffin wax block after removing tissue cores

Fig. 13.12 Tissue microarray blocks

Fig. 13.13 The number of probes on each tissue microarray slide can vary depending on the diameter of punched specimens

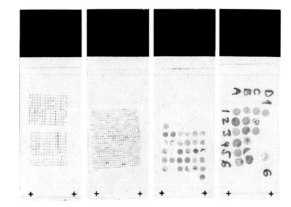

(Kononen et al. 1998) demonstrated the power of this technique for defining new subgroups of tumors.

Examples of potential applications for tissue microarrays include testing and optimization of probes and antibodies, the organization of large tissue repositories, and the facilitation of multicenter studies. Further, tissue microarrays can be used for educational purposes as well as to improve quality control and standardization of staining methods and interpretation (Moch et al. 2001). Tissue microarrays have become one of the most promising tools for the molecular and anatomic pathologist and will have many applications in cancer research and diagnostic pathology. Readers wanting further depths of knowledge in tissue microarray technology are referred to the following websites: http://tissuearray.org/; http://tmalab.jhmi.edu/; and http://www.protocol-online.org/prot/Histology/Tissue_Microarray__TMA_/).

References

Abd El-Rehim DM, Pinder SE, Paish CE, Bell J, Blamey RW, Robertson JF, Nicholson RI, Ellis IO (2004) Expression of luminal and basal cytokeratins in human breast carcinoma. J Pathol 203:661–671

Altmannsberger M, Dirk T, Droese M, Weber K, Osborn M (1986) Keratin polypeptide distribution in benign and malignant breast tumors: subdivision of ductal carcinomas using monoclonal antibodies. Virchows Arch B Cell Pathol Incl Mol Pathol 51(3):265–275

Assefnia S, Jones LP, Torre KM, Furth PA (2006) Expression of p63 in a mouse model of Brca1-mutation-related breast cancer. Poster presented at the Annual Meeting of the American Association for Cancer Research, Washington, DC (Meeting Abstracts)

Bártek J, Vojtěsek B, Stasková Z, Bártková J, Kerekés Z, Rejthar A, Kovarík J (1991) A series of 14 new monoclonal antibodies to keratins: characterization and value in diagnostic histopathology. J Pathol 164(3):215–224

Battifora H (1986) The multitumor (sausage) tissue block – novel method for immunohistochemical antibody testing. Lab Invest 55:244–248

Böcker W, Moll R, Poremba C, Holland R, Van Diest PJ, Dervan P, Burger H, Wai D, Ina DR, Brandt B, Herbst H, Schmidt A, Lerch MM, Buchwalow IB (2002) Common adult stem cells in the human breast give rise to glandular and myoepithelial cell lineages: a new cell biological concept. Lab Invest 82(6):737–746

Böcker W et al (2006) Preneoplasia of the breast: a new conceptual approach to proliferative breast disease. Elsevier, München

Chu PG, Weiss LM (2002a) Expression of cytokeratin 5/6 in epithelial neoplasms: an immunohistochemical study of 509 cases. Mod Pathol 15(1):6–10

Chu PG, Weiss LM (2002b) Keratin expression in human tissues and neoplasms. Histopathology 40(5):403–439

Coons AH, Creech H, Jones R (1941) Immunological properties of an antibody containing a fluorescent group. Proc Soc Exp Biol Med 47:200–202

Coons AH, Creech H, Jones R, Berliner E (1942) The demonstration of pneumococcal antigen in tissues by the use of fluorescent antibody. J Immunol 45:159–170

Dabbs DJ (ed) (2006) Diagnostic immunohistochemistry. Churchill Livingstone, Philadelphia

Dairkee SH, Mayall BH, Smith HS, Hackett AJ (1987) Monoclonal marker that predicts early recurrence of breast cancer. Lancet 1(8531):514

Demirkesen C, Hoede N, Moll R (1995) Epithelial markers and differentiation in adnexal neoplasms of the skin: an immunohistochemical study including individual cytokeratins. J Cutan Pathol 22(6):518–535

Diaz LK, Cryns VL, Symmans WF, Sneige N (2007) Triple negative breast carcinoma and the basal phenotype: from expression profiling to clinical practice. Adv Anat Pathol 14 (6):419–430

Fillies T, Werkmeister R, Packeisen J, Brandt B, Morin P, Weingart D, Joos U, Buerger H (2006) Cytokeratin 8/18 expression indicates a poor prognosis in squamous cell carcinomas of the oral cavity. BMC Cancer 6:10

Foschini MP, Scarpellini F, Gown AM (2000) Differential expression of myoepithelial markers in salivary, sweat and mammary glands. Int J Surg Pathol 8:29–37

Franke WW, Schiller DL, Moll R, Winter S, Schmid E, Engelbrecht I, Denk H, Krepler R, Platzer B (1981) Diversity of cytokeratins. Differentiation specific expression of cytokeratin polypeptides in epithelial cells and tissues. J Mol Biol 153(4):933–959

Fuchs E, Green H (1980) Changes in keratin gene expression during terminal differentiation of the keratinocyte. Cell 19(4):1033–1042

Gusterson BA, Ross DT, Heath VJ, Stein T (2005) Basal cytokeratins and their relationship to the cellular origin and functional classification of breast cancer. Breast Cancer Res 7(4):143–148

Hsu FD, Nielsen TO, Alkushi A, Dupuis B, Huntsman D, Liu CL, van de Rijn M, Gilks CB (2002) Tissue microarrays are an effective quality assurance tool for diagnostic immunohistochemistry. Mod Pathol 15:1374–1380

Ivanyi D, Ansink A, Groeneveld E, Hageman PC, Mooi WJ, Heintz AP (1989) New monoclonal antibodies recognizing epidermal differentiation-associated keratins in formalin-fixed, paraffin-embedded tissue. Keratin 10 expression in carcinoma of the vulva. J Pathol 159 (1):7–12

Jacquemier J, Padovani L, Rabayrol L, Lakhani SR, Penault-Llorca F, Denoux Y, Fiche M, Figueiro P, Maisongrosse V, Ledoussal V, Martinez PJ, Udvarhely N, El MG, Ginestier C, Geneix J, Charafe-Jauffret E, Xerri L, Eisinger F, Birnbaum D, Sobol H (2005) Typical medullary breast carcinomas have a basal/myoepithelial phenotype. J Pathol 207:260–268

Jetten AM, Harvat BL (1997) Epidermal differentiation and squamous metaplasia: from stem cell to cell death. J Dermatol 24:711–725

Kalof AN, Tam D, Beatty B, Cooper K (2004) Immunostaining patterns of myoepithelial cells in breast lesions: a comparison of CD10 and smooth muscle myosin heavy chain. J Clin Pathol 57:625–629

Kononen J, Bubendorf L, Kallionimeni A, Barlund M, Schraml P, Leighton S, Torhorst J, Mihatsch MJ, Sauter G, Kallionimeni OP (1998) Tissue microarrays for high-throughput molecular profiling of tumor specimens. Nat Med 4:844–847

Kraaz W, Risberg B, Hussein A (1988) Multiblock – an aid in diagnostic immunohistochemistry. J Clinic Pathol 41:1337

Lobeck H, Bartke I, Naujoks K, Müller D, Bornhöft G, Mischke D, Wild G (1989) Vertei-lungsmuster der Zytokeratinpolypeptide 4 und 5 im normalen und neoplastischen Epithel unter Verwendung neuer paraffingängiger monoklonaler Antikörper (eine immunhistochem-ische Untersuchung). Verh Dtsch Ges Pathol 73:645

Malzahn K, Mitze M, Thoenes M, Moll R (1998) Biological and prognostic significance of stratified epithelial cytokeratins in infiltrating ductal breast carcinomas. Virchows Arch 433 (2):119–129

Moch H, Kononen J, Kallioniemi OP, Sauter G (2001) Tissue microarrays: what will they bring to molecular and anatomic pathology? Adv Anat Pathol 8:14–20

Moll R (1998) Cytokeratins as markers of differentiation in the diagnosis of epithelial tumors. Subcell Biochem 31:205–262

Moll R, Franke WW, Schiller DL, Geiger B, Krepler R (1982a) The catalog of human cytoker-atins: patterns of expression in normal epithelia, tumors and cultured cells. Cell 31(1):11–24

Moll R, Franke WW, Volc-Platzer B, Krepler R (1982b) Different keratin polypeptides in epidermis and other epithelia of human skin: a specific cytokeratin of molecular weight 46,000 in epithelia of the pilosebaceous tract and basal cell epitheliomas. J Cell Biol 95(1):285–295

Moll R, Krepler R, Franke WW (1983) Complex cytokeratin polypeptide patterns observed in certain human carcinomas. Differentiation 23(3):256–269

Moll R, Dhouailly D, Sun TT (1989) Expression of keratin 5 as a distinctive feature of epithelial and biphasic mesotheliomas. An immunohistochemical study using monoclonal antibody AE14. Virchows Arch B Cell Pathol Incl Mol Pathol 58(2):129–145

Moll R, Schiller DL, Franke WW (1990) Identification of protein IT of the intestinal cytoskeleton as a novel type I cytokeratin with unusual properties and expression patterns. J Cell Biol 111(2):567–580

Moll R, Löwe A, Laufer J, Franke WW (1992) Cytokeratin 20 in human carcinomas. A new histodiagnostic marker detected by monoclonal antibodies. Am J Pathol 140:427–447

Moll R, Divo M, Langbein L (2008) The human keratins: biology and pathology. Histochem Cell Biol 129:705–733

Moritani S, Kushima R, Sugihara H et al (2002) Availability of CD10 immunohistochemistry as a marker of breast myoepithelial cells on paraffin sections. Mod Pathol 15:397–405

Narahashi T, Niki T, Wang T, Goto A, Matsubara D, Funata N, Fukayama M (2006) Cytoplasmic localization of p63 is associated with poor patient survival in lung adenocarcinoma. Histopathology 49(4):349–357

Otterbach F, Bankfalvi A, Bergner S, Decker T, Krech R, Böcker W (2000) Cytokeratin 5/6 immunohistochemistry assists the differential diagnosis of atypical proliferations of the breast. Histopathology 37(3):232–240

Owens DW, Lane EB (2003) The quest for the function of simple epithelial keratins. Bioessays 25(8):748–758

Pia-Foschini M, Reis-Filho JS, Eusebi V, Lakhani SR (2003) Salivary gland-like tumours of the breast: surgical and molecular pathology. J Clin Pathol 56:497–506

Purkis PE, Steel JB, Mackenzie IC, Nathrath WB, Leigh IM, Lane EB (1990) Antibody markers of basal cells in complex epithelia. J Cell Sci 97(Pt 1):39–50

Reis-Filho JS, Tutt AN (2008) Triple negative tumours: a critical review. Histopathology 52:108–118

Schweizer J, Bowden PE, Coulombe PA, Langbein L, Lane EB, Magin TM, Maltais L, Omary MB, Parry DA, Rogers MA, Wright MW (2006) New consensus nomenclature for mammalian keratins. J Cell Biol 174(2):169–174

Sørlie T, Perou CM, Tibshirani R, Aas T, Geisler S, Johnsen H, Hastie T, Eisen MB, van de RM, Jeffrey SS, Thorsen T, Quist H, Matese JC, Brown PO, Botstein D, Eystein LP, Borresen-Dale AL (2001) Gene expression patterns of breast carcinomas distinguish tumor subclasses with clinical implications. Proc Natl Acad Sci USA 98(19):10869–10874

Troyanovsky SM, Guelstein VI, Tchipysheva TA, Krutovskikh VA, Bannikov GA (1989) Patterns of expression of keratin 17 in human epithelia: dependency on cell position. J Cell Sci 93 (Pt 3):419–426

Troyanovsky SM, Leube RE, Franke WW (1992) Characterization of the human gene encoding cytokeratin 17 and its expression pattern. Eur J Cell Biol 59(1):127–137

van de Rijn RM, Perou CM, Tibshirani R, Haas P, Kallioniemi O, Kononen J, Torhorst J, Sauter G, Zuber M, Kochli OR, Mross F, Dieterich H, Seitz R, Ross D, Botstein D, Brown P (2002) Expression of cytokeratins 17 and 5 identifies a group of breast carcinomas with poor clinical outcome. Am J Pathol 161(6):1991–1996

Wetzels RH, Kuijpers HJ, Lane EB, Leigh IM, Troyanovsky SM, Holland R, van Haelst UJ, Ramaekers FC (1991) Basal cell-specific and hyperproliferation-related keratins in human breast cancer. Am J Pathol 138(3):751–763

Yang A, Kaghad M, Wang Y, Gillett E, Fleming MD, Dötsch V et al (1998) p63, a p53 homolog at 3q27-29, encodes multiple products with transactivating, death-inducing, and dominant-negative activities. Mol Cell 2:305–316

Chapter 14
A Picture Is Worth a Thousand Words

Fully realizing that immunohistochemistry is a visual art, we therefore need optical aids to assess the results of immunohistochemical reaction. The major optical instrument available to the immunohistochemist is the light microscope — probably the most well-known and well-used research tool in biology and medicine. Yet many students and researchers are unaware of the full range of features that are available in light microscopes. Severe limitations in exploiting the light microscope can cause considerable confusion in interpretation of immunohistochemical stainings. In this chapter we attempt to address this issue without the unnecessary math that often confuses students and postdoctoral researchers. For beginners we highly recommend the handbook "Light Microscopy, Essential Data" by Rubbi (1994) and the review by Davidson and Abramowitz "Optical microscopy" available on the website: http://micro.magnet.fsu.edu/primer/pdfs/microscopy.pdf. Explanations of the common words used in microscopy may be found in the "RMS Dictionary of Light Microscopy" (Bradbury et al. 1989) and on the website: http://www.microscope-microscope.org/basic/microscope-glossary.htm.

14.1 Brightfield Microscopy

Brightfield microscopy is the simplest of all the light microscopy techniques using transmitted light illumination. A basic light microscope has a built-in illuminator, adjustable condenser, mechanical stage, objective, binocular eyepiece tube, and camera (Fig. 14.1). The purpose of the condenser is to concentrate the light onto the specimen; its aperture diaphragm regulates resolution and contrast. With Kohler illumination (see below), the condenser focuses the image of the field diaphragm located in front of the collecting lens of the illuminator onto the image plane. After passing through the specimen and the system of lenses (condenser, objective and ocular), the light displays a magnified view of the sample. The magnification of the image is simply the objective lens magnification (usually stamped on the lens body) times the ocular magnification. The maximum useful magnification for light

I.B. Buchwalow and W. Böcker, *Immunohistochemistry: Basics and Methods*,
DOI 10.1007/978-3-642-04609-4_14, © Springer-Verlag Berlin Heidelberg 2010

Fig. 14.1 Principal
components of a light
microscope. Adapted from
Rubbi (1994)

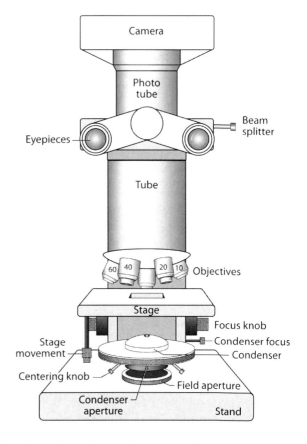

microscopes is around 1000×. Everything above this value will result in "empty magnification", that is, magnification without further detail. Conventional light microscopes are theoretically capable of resolving structures as close as 0.2 μm. The reason for this limit lies in the physical nature of light; it is not possible to resolve details that are smaller than the wavelength of the light used. The resulting image can be detected directly by the eye, imaged on a photographic film or captured digitally. The most recent development is the digital microscope which uses a charge-coupled device (CCD) camera. The CCD camera is attached to a computer via a USB or fire-wire port, and the image is shown on the computer screen. As digital technology has become cheaper in recent decades, it has replaced the old film methods for most purposes.

14.1.1 How to Set Up a Light Microscope Properly for Transmitted Light Illumination: Köhler Illumination

Histochemists are usually aware of the use of focus knobs, used to sharpen the image of the specimen, but are frequently unaware of adjustments to the condenser that can affect resolution and contrast. However, there is little point in having an expensive optical system if it is not producing the quality images of which it is capable – this is the equivalent of running a Mercedes with a one-horsepower motor. In order to get the best image possible with the light microscope, it is crucial that the condenser be set up properly according to the technique invented by a German scientist August Köhler in 1893. This technique provides bright, even illumination and is still in general use today. Köhler illumination is the method of choice for the majority of modern microscopes. It provides optimum contrast and resolution by focusing and centering the light path and spreading the light evenly over the field of view. For how to set up Köhler illumination see Fig. 14.2.

Upon setting up Köhler illumination you can adjust contrast using the condenser aperture diaphragm. The condenser aperture diaphragm controls the angle of the light beam coming up through the condenser. When the diaphragm is wide open, the image is brighter and contrast is low. When the diaphragm is nearly closed, the light comes straight up through the centre of the condenser lens and contrast is high. The amount of contrast added will also depend on the sample. Be careful when adjusting the condenser diaphragm, since closing the condenser diaphragm reduces resolution. Moreover, too much contrast can introduce artifacts into your images. To maximize both contrast and resolution, close the diaphragm just to the point where the image begins to get dark and no further. Never use condenser aperture to control light intensity. Illumination intensity can be varied by adjusting the voltage to the light source. Adjust illumination so that the field is bright without hurting the eyes.

14.1.2   The Choice of Microscope Objectives

The most important components of the light microscope are objectives because they are responsible for primary image formation and play a central role in determining the quality of images that the microscope is capable of producing. Yet few researchers grasp the differences between specific objective classes. Objectives are made with different degrees of correction for spherical and chromatic aberration, field size and flatness, transmission wavelengths, and other factors. Chromatic aberration leads to problems, with different wavelengths of light being focused to different positions within your sample. Spherical aberration results in inaccurate focusing of light due to the curved surface of the lens, whereby light rays passing through the lens at different distances from its center are focused to different positions in the Z-axis. It is the major cause of loss of signal intensity and resolution with increasing focus depth through thick specimens.

The least expensive (and most common) objectives are the *achromatic objectives* that are designed to limit the effects of chromatic and spherical aberration. Achromatic objectives are corrected to bring two wavelengths of light (typically red and blue) into focus in the same plane. The limited correction of achromatic objectives leads to problems with color microscopy and photomicrography.

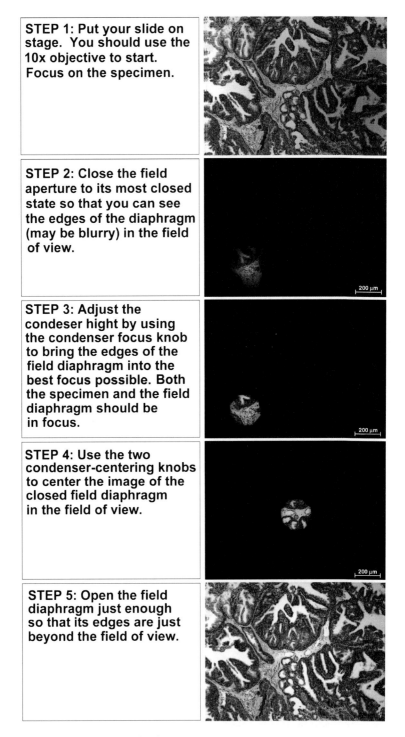

STEP 1: Put your slide on stage. You should use the 10x objective to start. Focus on the specimen.

STEP 2: Close the field aperture to its most closed state so that you can see the edges of the diaphragm (may be blurry) in the field of view.

STEP 3: Adjust the condeser hight by using the condenser focus knob to bring the edges of the field diaphragm into the best focus possible. Both the specimen and the field diaphragm should be in focus.

STEP 4: Use the two condenser-centering knobs to center the image of the closed field diaphragm in the field of view.

STEP 5: Open the field diaphragm just enough so that its edges are just beyond the field of view.

Fig. 14.2 Setting up Köhler illumination

A higher level of correction (and expense) is found in *apochromatic objectives*, which are corrected chromatically for three colors (red, green, and blue), almost eliminating chromatic aberration, and are corrected spherically for two colors. Apochromatic objectives are a good choice for color photomicrography in white light (http://micro.magnet.fsu.edu/primer/pdfs/microscopy.pdf). *Fluorite objectives* are chromatically better corrected than the achromatic objectives, but worse than the plan-apochromat (see below) objectives. The contrast of the fluorite objectives is however larger than with the plan-apochromat objectives.

The "king class" under the objectives are *plan-apochromat* and *plan-fluorite objectives* designed to correct all optical aberrations throughout the visible spectrum from violet to red from center to edges across the entire field of view. They deliver superior image flatness and color reproduction, and resolve power at the theoretical limit of today's optical technology. These objectives are the perfect choice for multistained fluorescence specimens and when using all transmitted-light methods. High transmission of wavelengths up to the near UV range (360 nm) and the use of special glass types of low autofluorescence in plan-fluorite objectives make the latter especially appropriate for fluorescence microscopy. Plan-fluorite objectives are the most versatile family of objectives that can be used for all common brightfield and fluorescence microscopy methods.

Because of their high level of correction, apochromat objectives usually have, for a given magnification, higher numerical apertures (NA) than do achromats. An aperture is the area of a lens that is available for the passage of light. The concept of the NA was introduced by Ernst Abbe, and developed over 130 years at Carl Zeiss. NA (usually stamped on the lens body) is related to the angular aperture of the lens and the index of refraction of the medium found between the lens and the specimen. Derived by a complex mathematical formula, NA determines the resolving power of the objective. Thus, a 40×1.3 NA objective lens will be able to resolve far finer details than a 40×0.75 NA lens, despite their similar magnification (North 2006). To get the best possible image in transmitted light, you should have a condenser system that matches or exceeds the NA of the highest power objective lens on your microscope. Care must be taken to guarantee that the condenser aperture is opened to the correct position with respect to the objective NA. The condenser aperture adjustment and proper focusing of the condenser are of critical importance in realizing the full potential of the objective. Be aware of the proper matching of all optical components. Note that combining objectives and microscopes from different manufacturers can introduce aberrations. It must also be noted that no objective is ideal for all circumstances. For objectives selection search and detailed description of classes of objectives, the reader is referred to the website http://www. zeiss.de/objectives and to reviews of Abramowitz et al. (2002) and Lichtman and Conchello (2005).

Along with the choice of the right objective, clean microscope optics are prerequisite for perfect images. Always make sure the stage and lenses are clean. Be careful, however, when cleaning microscope optics, since every cleaning process involves the risk of scratching the optics (or doing other damage). To clean optical surfaces of objectives, oculars and condenser, use an appropriate lens cleaner or distilled water to help remove dried material. If you use the wrong cleaning solvent, you run the risk of removing the optical coating of the lenses, and/or of softening the lens kit which holds the lens in place. Never use paper towels, your shirt, or any material other than good quality lens cleaning tissue or a cotton swab. The choice of the best cleaning method depends on the nature of the optical surface and the type of dirt to be removed. For more information to this issue, see the website http://www.zeiss.com/cleanmicroscope. Cover the instrument with a dust jacket when not in use.

14.2   Fluorescence Microscopy

The birth of fluorescence microscopy took place at Carl Zeiss thanks to August Köhler, when in 1904 he described a phenomenon of fluorescence of biological objects under UV-illumination. The fluorescence microscopy exploits the property of certain materials to emit energy detectable as visible light when irradiated with the light of a specific wavelength. Understanding the principles underlying fluorescence microscopy is extremely important for solving imaging problems and adequately assessing the results of immunofluorescent stainings.

The fluorescence microscope generates the image by a completely different way from the normal brightfield microscope. The excitatory light is passed from above (or, for inverted microscopes, from below), through the objective and then onto the specimen instead of passing it first through the specimen. In other words, the microscope objective not only has the familiar role of imaging and magnifying the specimen, but also serves as the condenser that illuminates it. Typical components of a modern fluorescence microscope are the light source (Xenon or Mercury arc-discharge lamp), the excitation filter, the dichroic mirror, and the emission filter (see Fig. 14.3).

14.2.1   The Choice of Filter Cubes for Fluorescence Microscopy

In order to observe fluorescence through the microscope it is necessary to illuminate the object with light of high intensity and of specific wavelengths, generally in the region between 300 nm and 700 nm. Thereby, the fluorophore absorbs light of specific wavelengths (excitation) and simultaneously re-emits part of this energy at longer wavelengths (emission), usually in the visible region of the spectrum. The difference between the absorption maximum and emission maximum is the Stokes' shift. British scientist Sir George G. Stokes first described this phenomenon in 1852. He noted that fluorescence emission always occurred at a longer wavelength than that of the excitation light. Stokes' shift is fundamental to fluorescence techniques. For example, fluorescein isothiocyanate (FITC) is excited by blue

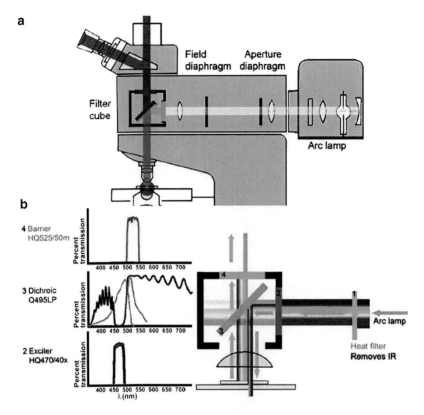

Fig. 14.3 The fluorescence microscope. **a** Epi-illumination fluorescence microscopes use the objective both to illuminate and image the specimen. Shown is an upright microscope with the slide at the bottom. The light source, in this case an arc lamp, sends full-spectrum light to the specimen by way of a fluorescence "cube" that selectively illuminates the specimen with a wavelengths that excite a particular fluorophore (shown, green light to excite rhodamine). The red fluorescence that is excited sends photons in all directions, and a fraction are collected by the objective and sent through the cube to the eye or camera port above. The cube has two filters, the dichroic mirror and barrier filter, to prevent the exciting wavelengths from reaching the detector. **b** The details of a cube designed by Chroma Technologies to excite and detect EGFP. The three main components (labeled *2, 3* and *4*) have specific spectral features that are ideal for GFP. Note that the dichroic mirror (*3*) splits between reflection and transmission right between the absorption and emission peaks of the GFP, which are superimposed in *blue* and *green*, respectively. Reprinted by permission from Macmillan Publishers Ltd: Nature Methods, Lichtman and Conchello (2005)

light and emits green light; rhodamine isothiocyanate is excited by green light and emits red light.

Fluorophores can be visualized in fluorescence microscopy using special filter blocks that are composed of the excitation filter, dichroic mirror and emission filter. The excitation filter must select wavelengths of light from a light source that fall in the maximum absorption region of the fluorophore. The emission filter must pass the fluorescent wavelengths but not the excitation wavelengths. The dichroic mirror (or so-called chromatic beam-splitter) reflects wavelengths of light below the transition wavelength value and transmits wavelengths above this value. This property accounts for the name given to this mirror (dichroic, two color). The most commonly used mirrors and filters are the interference ones. They consist of layers of materials with different refractive indices and thicknesses. The dichroic mirror of the interference type for fluorescence microscopy was first made by a Russian scientist Evgenii Brumberg in St Petersburg in the early 1950s (Brumberg and Krylova 1953). Since then, filters and mirrors of this type have been the key elements of filter sets used in fluorescence microscopy. In modern fluorescence microscopes, manufacturers mount such dichroic mirrors together with excitation and emission filters on an optical block commonly referred to as a filter cube (http://microscope.fsu.edu/primer/techniques/fluorescence/filters.html). In the filter cube, filters are located normal to the optic axis, while chromatic beam splitters usually work at 45°. Such cubes fit into a circular carousel or linear block that can hold from three to nine separate cubes. These cubes can be moved into position either manually or by a computer driven motor.

In addition to the standard fluorescence cubes, some manufacturers offer a cube (IGS Cube) for immuno-gold (or silver) staining. This cube, in place of a dichroic mirror, has a standard half-mirror similar to the kind used in metallurgical bright-field reflected light microscopy. For a proper selection of filter cubes matching the spectral characteristics of the fluorophore the reader is referred to the review article "Fluorescence microscopy" by Lichtman and Conchello (2005) covering all of the important aspects necessary to understand the basic principles of fluorescence, the characteristics of fluorophores, excitation and emission spectral properties and microscope parameters, as well as to the following websites: http://www.zeiss.de/objectives; http://microscope.fsu.edu/primer/techniques/fluorescence/filters.html; http://www.jacksonimmuno.com/home.asp; https://www.omegafilters.com/; http://micro.magnet.fsu.edu/primer/java/fluorescence/photobleaching/.

The choice of filter cubes for fluorescence microscopy must match the spectral characteristics (excitation and emission) of the fluorophore (see Table 14.1) used to label the specimen. Long-pass filter sets, collecting all emissions past a certain wavelength, and broad-band cubes give a stronger signal for a given fluorophore than narrow-band cubes and therefore may be preferred when one is looking for a signal from a single dye, for instance Cy5, Cy3 or Alexa Fluor 488. For multiple labeling, it is critical to use narrow-band cubes, which collect emissions in a specific range, thus avoiding overlapping in fluorescence between fluorophores; the narrower the range of the band-pass filter, the better it can separate fluoro-chromes with close emission spectra.

The standard fluorescence microscope can be equipped with filter sets for observation of the four types of fluorescent label that are in common use:

1. Filter set for visualization of Cy 5 or Alexa Fluor 647.
2. Filter set for visualization of Cy 3 or RRX.
3. Filter set for visualization of GFP, FITC or Alexa Fluor 488.
4. Filter set for visualization of AMCA, Alexa Fluor 350 or nuclear dye DAPI.

Table 14.1 Characteristics of some fluorophores used for fluorescence microscopy in order of excitation wavelength. A more extended list of fluorophores can be found on the website: http://flowcyt.salk.edu/fluo.html

Fluorophores for antibody labeling	Excitation wavelength (nm)	Emission wavelength (nm)	Color
Alexa Fluor 350	346	445	Blue
AMCA (7-amino-4-methylcoumarin-3-acetic acid)	350	450	Blue
DyLight 488	493	518	Green
Alexa Fluor 488	494	517	Green
FITC (Fluorescein-isothiocyanate)	490	520	Green
Oyster 488	502	523	Green
Cy3 (Cyanine 3)	552	565	Red
Oyster 555	552	572	Red
Alexa Fluor 555	556	573	Red
Rhodamine Red-X (RRX)	570	590	Red
Alexa Fluor 568	578	603	Red
Alexa Fluor 594	594	615	Red
TR (Texas Red)	596	620	Red
Cy5[a] (Cyanine 5)	650	667	Deep red
Oyster 647	651	671	Deep red
DyLight 649	654	673	Deep red
Alexa Fluor 647[a]	650	668	Deep red
Alexa Fluor 700[a]	702	723	Deep red

[a]Human vision is insensitive to light beyond ~650 nm, and therefore it is not possible to view the far-red fluorescent dyes by looking through the eyepiece of a conventional fluorescence microscope. However, the fluorescence in this spectrum region is efficiently registered by sensitive digital cameras

14.2.2 The Choice of Fluorophores for Multiple Immunostaining

Albert H. Coons was the first to attach a fluorescent dye (fluorescein isocyanate) to an antibody and to use this antibody to localize its respective antigen in a tissue section. Fluorescein, one of the most popular fluorochromes ever designed, has enjoyed extensive application in immunofluorescence labeling. For many years, classical fluorescent probes such as FITC or Texas red (TR) have been successfully utilized in fluorescence microscopy. In recent decades, brighter and more stable fluorochromes have continually been developed (see Table 14.1). Marketed by a number of distributors, cyanine dyes, Cy2, Cy3, Cy5, Cy7, feature enhanced water solubility and photostability as well as a higher fluorescence emission intensity as compared to many of the traditional dyes, such as FITC or TR.

Dramatic advances in modern fluorophore technology have been achieved with the introduction of Alexa Fluor dyes by Molecular Probes (Alexa Fluor is a registered trademark of Molecular Probes). Alexa Fluor dyes are available in a broad range of fluorescence excitation and emission wavelength maxima, ranging from the ultraviolet and deep blue to the near-infrared regions. Because of the large number of available excitation and emission wavelengths in the Alexa Fluor series, multiple labeling experiments can often be conducted exclusively with these dyes. Emission profiles of Alexa Fluor series are comparable in spectral width to the cyanine dyes (http://www.olympusconfocal.com/theory/fluorophoresintro.html). More recently, Thermo Fisher Scientific Inc. (http://www.piercenet.com/) have introduced for use in fluorescence microscopy a new family of nine DyLight fluorescent dyes featuring a high fluorescence emission intensity and strong photostability.

Another promising approach was achieved by Luminartis. Based on a proprietary technology, fluorescent dyes with excellent water solubility, high fluorescence emission and photostability are provided. The improved biocompatibility gives rise to decreased background and concomitantly increased sensitivity of the resulting conjugates. Moreover, depending on the antibody used, high degrees of labeling may be realized, without observing quenching effects. The fluorescent dyes, which cover a spectral absorption range from 502 to 783 nm, are commercially available as Oyster dyes (www.luminartis.com).

For multiple immunolabeling in colocalization studies, it is important to use correct filter sets and choose fluorophores with well-separated excitation and emission spectra (http://www.olympusfluoview.com/applications/colocalization. html). For instance, Alexa Fluor 488 or FITC (490–520 nm) can be used with Cy3 (552–565 nm) or RRX (570–590 nm) for double labeling. Thereby, RRX might be a better choice in this fluorophore class due to its higher emission maximum (at 590 nm), excluding possible emission spectral overlap with fluorescein when using broad-band filter cubes. As a third label, far-red fluorescent dyes, such as Cy5, DyLight 649 or Alexa Fluor 647 (650–670 nm), can be exploited successfully (see Table 14.1). Because of their far-red emission maximum, far-red fluorescent dyes cannot be seen well by eye but they are often utilized as a third fluorophore in triple labeling experiments due to a wide separation of their emission from that of shorter-wavelength-emitting fluorophores. Another significant advantage of such dyes over other fluorophores is that a wavelength used for excitation is out of the range of natural autofluorescence of biological specimens. This makes it possible to effectively avoid tissue autofluorescence, which is annoying only at shorter wavelengths (see Sect. 5.5). By attaching different fluorophores to different antibodies, the distribution of two or more antigens can be determined, in contrast to brightfield microscopy, even in the same cell and in the same subcellular compartments (see Sect. 8).

For readers wanting further depth of knowledge with regard to the use of fluorescence and brightfield microscopy methods, several hundred books dealing with various aspects of optical microscopy and related fields are currently available from the booksellers (see website: http://www.vanosta.be/pcrref.htm). The top ten recommended books on microscopy, digital imaging, fluorescence and microtechnique are listed on the Molecular Expressions website: http://micro.magnet.fsu.edu/primer/books/index.html.

References

Abramowitz M, Spring KR, Keller HE, Davidson MW (2002) Basic principles of microscope objectives. Biotechniques 33:772–781

Bradbury S, Evenett PJ, Haselmann H, Piller H (1989) RMS Dictionary of Light Microscopy (Royal Microscopical Society/Microscopy Handbook), vol 15. Oxford University Press, Oxford, UK

Brumberg EM, Krylova TN (1953) Application dividing mirrors for interferometry in fluorescent microscopy. Zh Obshch Biol 14:461 464

Davidson MW, Abramowitz M. Optical microscopy. http://micro.magnet.fsu.edu/primer/pdfs/microscopy.pdf

Lichtman JF, Conchello J-A (2005) Fluorescence microscopy. Nat Methods 2:910–919

North AJ (2006) Seeing is believing? A beginners' guide to practical pitfalls in image acquisition. J Cell Biol 172:9–18

Rubbi CP (1994) Light Microscopy, Essential Data (Essential Data Series). Wiley, New York

Glossary

Absorbed antibody See "Preaxbsorbed antibody."

Absorption or preabsorption control a control, in which the primary antibody (prior to its use) is incubated for 1 hr with a tenfold molar excess of the purified antigen. Absent or greatly diminished immunostaining should be obtained after application of this preabsorbed antibody.

Achromat Microscope objective that is designed to limit the effects of chromatic and spherical aberration. Achromatic objectives are corrected to bring two wavelengths of light (typically red and blue) into focus in the same plane.

Affinity-purified antibodies are isolated from antisera by immunoaffinity chromatography using antigens coupled to agarose beads.

Alkaline phosphatase (AP) an enzyme routinely used in immunohistochemistry for labeling antibodies. AP catalyzes the hydrolysis of a variety of phosphate–containing substances in the alkaline pH range. The enzymatic activity of alkaline phosphatase can be localized by coupling a soluble product generated during the hydrolytic reaction with a "capture reagent," producing a colored insoluble precipitate.

Angstrom (symbol Å) is equal to 0.1 nm or 1×10^{-10}m.

Antifade agents (free-radical scavengers) in aqueous mounting media prevent photobleaching of fluorophore labels under exposure to excitation light.

Antiserum (plural: antisera) Blood serum containing one or more polyclonal antibodies that are specific for one or more antigens. The antibodies in an antiserum result from previous immunization or exposure to an agent of disease.

Antibody Antibodies (also known as immunoglobulins) are proteins that are found in blood or other bodily fluids of vertebrates, and are used by the immune system to identify and neutralize foreign objects, such as bacteria and viruses. Production of antibodies is the main function of the humoral immune system. Several immunoassay methods including immunohistochemistry are based on detection of complex antigen–antibody.

Antibody affinity from the Latin, *affinis* = *connected with*, having things in common. In immunohistochemistry, antibody affinity determines the strength of binding of a monovalent antibody, such as Fab fragment, to one epitope, i.e., how tightly an antibody binds to its particular antigen.

Antibody avidity is commonly applied to antigen–antibody interaction, where multiple, weak, noncovalent bonds form between antigen and antibody. Avidity is distinct from affinity, which is a term used to describe the strength of a single bond. As such, avidity is the combined synergistic strength of bond affinities rather than the sum of bonds.

Antibody classes In mammals, there are five classes of antibodies (also known as immunoglobulins: IgG, IgA, IgM, IgE and IgD). Each immunoglobulin class differs in their heavy chain constant domains (see Fig. 1.2) and consequently in its biological properties. For immunoassays, two immuno-glubulin classes are of importance – IgG and IgM.

Antibody isotypes Differences in the amino acid sequence and disulfide bonding patterns of the heavy chain in antibody classes give rise to antibody subclasses (also referred to as isotypes). For instance, mouse IgG antibodies are divided into four isotypes: IgG1, IgG2, IgG2a, and IgG3.

Antigen is a substance that stimulates an immune response, especially the production of antibodies. Antigens are usually proteins or polysaccharides, but can be any type of molecule, including small molecules (haptens) coupled to a carrier-protein.

Antigen retrieval (antigen unmasking or epitope unmasking) is predominantly defined as a high-temperature heating method. To recover the antigenicity of tissue sections that had been masked by formalin fixation and paraffin imbedding and to enhance the signal of immunohistochemistry, archival paraffin-embedded tissue sections are subjected to boiling in buffered water solutions.

Aperture is the area of a lens that is available for the passage of light.

Apochromat Microscope objective that has better color correction than the much more common achromat objectives corrected chromatically for two wavelengths of light (red and blue). Apochromatic objectives are corrected chromatically for three colors (red, green and blue) and spherically for two colors, which practically eliminates chromatic aberration.

Autofluorescence or primary fluorescence Fluorescence from objects in a microscope sample other than from fluorophores. In mammalian tissues, natural fluorescence is due in large part to substances like flavins and porphyrins, lipofuscins (the breakdown product of old red blood cells – an "aging" pigment; also prominent in certain large neurons in the CNS), elastin and collagen (connective tissue components, particularly in blood vessel walls) and some others. Because of its broad excitation and emission spectra, autofluorescence overlaps with working spectra of most commonly used fluorophores.

Avidin was first isolated from chicken egg white by Esmond Emerson Snell (1914–2003). This tetrameric protein contains four identical subunits (homotetramer) each of which can bind to biotin (Vitamin B_7, vitamin H) with a high degree of affinity and specificity. When these two molecules are in the same solution, they will bind with such high affinity that the binding is essentially irreversible. No matter how many times you wash them, they will not let go of each other.

Avidin–biotin complex (ABC) is based on the high affinity that streptavidin (from *Streptomyces avidinii*) and avidin (from chicken egg) have for biotin. Biotin is a naturally occurring vitamin. One mole avidin will bind four moles biotin. ABC method affords a several-fold higher antigen detectability than those achieved in the standard indirect method.

Background staining In most cases, background staining is not caused by a single factor. Along with Fc receptors, frequent causes of background staining are endogenous enzyme activity, if you use peroxidase or alkaline phosphatase as enzyme markers on your secondary antibodies, and endogenous biotin when using a streptavidin or avidin reagents. When fluorescent dyes are used in the experiments, autofluorescence (or natural fluorescence) of some tissue components can cause background problems and complicate the use of fluorescence microscopy.

Chain polymer-conjugated technology Developed by DakoCytomation, this technology utilizes an enzyme-labeled inert "spine" molecule of dextran. In addition to an average of 70 molecules of enzyme (AP or HRP), 10 molecules of antibody can be attached to the spine molecule. This technology allows a several-fold higher antigen detectability than those achieved in standard immunostaining protocols.

Chromatic aberration is the phenomenon of different wavelengths of light being focused to different positions within your sample.

Colloidal gold A suspension (or colloid) of submicrometer-sized particles of gold in a fluid, usually water. A colloidal gold conjugate consists of gold particles coated with a selected protein or macromolecule, such as an antibody, *protein A or protein G*. Because of their high electron density, the gold particles are visible in the electron microscope without further treatment.

Diaminobenzidine 3,3'-Diaminobenzidine (DAB) is an organic compound. As its water-soluble tetrahydrochloride derivative, it is used in enzymohisto-chemical and immunohistochemical stainings. DAB is oxidized by hydrogen peroxide in the presence of peroxidase to give a dark brown color. Be aware that DAB is harmful by inhalation, ingestion and if absorbed through the skin. It is eye, skin and respiratory irritant. May act as a carcinogen.

Dichroic mirror (or so called chromatic beam-splitter) reflects wavelengths of light below the transition wavelength value and transmits wavelengths above this value.

Direct and indirect immunostaining methods The *direct method* is a one-step staining method, and involves a labeled antibody reacting directly with the antigen in tissue sections. In this method, the primary antibody can be labeled with a fluorophore or biotin. In *indirect immunostaining*, the bound unlabeled primary antibody (first layer) is visualized with a secondary antibody (second layer) bearing label, such as a fluorophore, biotin or an enzyme.

Enzyme is a globular protein with an active site which binds to substrate molecules and helps to catalyze a reaction by holding melecules in the correct spatial conformation for the reaction to take place. Enzymes are important to health; many diseases are derived fron the lack of enzymes in the body. In enzyme-histochemical chromogenic reactions, a soluble colorless substrate is converted into a water-insoluble colored compound either directly or in a coupled reaction. In immunohistochemistry, enzyme labels are usually coupled to antibodies or to (strept)avidin.

Epitope is an antigenic determinant (a part of a molecule to which an antibody binds), which is a biological structure or sequence, such as a protein or carbohydrate, that is recognized by the immune system and to which an antibody binds.

Epitope tags In most cases, epitope tags are constructed of amino acids (from 6 to 15 amino acids long) and added to a molecule (usually proteins) which an investigator wants to visualize. The gene from the target protein is inserted into the epitope tag vector and the target protein with its tag is expressed in cells by transfection of the vector. When using invisible epitope tags (e.g., *c-myc* – a 10-amino acid segment of the human protooncogene myc; *HA* – hemagglutinin protein from human influenza hemagglutinin protein; *HIS* – six histidines placed in a row), the target proteins can be visualized by immunohistochemical procedures with anti-tag antibodies. Fluorescent epitope tags, such as green fluorescent protein (*GFP*) isolated from the jellyfish *Aequorea victoria* or red fluorescent proteins isolated from other species, including coral reef organisms, can be observed directly using fluorescence microscopy.

Emission filter must pass the fluorescent wavelengths but not the excitation wavelengths.

Enzymatic markers used in immunohistochemistry: Horseradish peroxidase (HRP) and calf intestinal or *E.coli* alkaline phosphatase (AP). Glucose oxidase from *Aspergillus niger* and *E.coli* β-galactosidase are only rarely applied.

Excitation filter selects wavelengths of light from a light source that fall in the maximum absorption region of a specific fluorophore.

Fab fragment (for *Fragment, antigen binding*) Antigen specific Fab "arms" of an antibody (IgG molecule) are responsible for antigen binding. Fab fragment is comprised of one light chain and the segment of heavy chain on the N-terminal side. The light chain and heavy chain segments are linked by interchain disulfide bonds.

Fc fragment (for *Fragment, crystallizable*) A region of an antibody (IgG molecule) composed of two heavy chains on the C-terminal side. Fc has many effector functions (e.g., binding complement, binding to cell receptors on macrophages and monocytes, etc.) and serves to distinguish one class of antibody from another.

Fc receptors are present on the cell membrane of macrophages, monocytes, granulocytes, lymphocytes and some other cells. They may, in theory, nonspecifically bind Fc portion of antibodies used for immunolabeling.

Ferritin is a globular protein complex consisting of 24 protein subunits and is the main intracellular iron storage protein in both prokaryotes and eukaryotes. Ferritin is used for immunolabeling at the electron microscope level because of its distinctive shape of the crystals and their electron density.

Fluorescence An optical phenomenon in which absorption of light by fluorescent molecules called fluorescent dyes or fluorophores is followed by the emission of light at longer wavelengths.

Fluorescence microscopes used in immunohistochemistry are so called epifluorescence microscopes (i.e., excitation of the fluorescence and observation are from above (epi) the specimen).

Fluorophore absorbs ultraviolet light (or violet, blue or green) and emits light of longer wavelength. Fluorophores are used in immunohistochemistry for labeling primary or secondary antibodies in direct and indirect immunostaining methods, respectively. They can be visualized in fluorescence microscopy using special filter sets.

Formaldehyde is the simplest aldehyde. Its chemical formula is H_2CO.

Formalin is commercial concentrated (37–40%) solution of formaldehyde. Formaldehyde fixative is made up of formalin diluted to a 10% solution (3.7–4% formaldehyde). Formaldehyde dissolved in phosphate buffered saline can be recommended as a universal fixative for the routine use.

Green fluorescent protein (GFP) was isolated from the jellyfish *Aequorea victoria*. Using biotechnology methods, GFP gene is fused with a host gene of interest and this chimera is transfected into the host genome. The resulting fusion protein that the cell produces is fluorescent. Other flourescent proteins, red fluorescent proteins, have been isolated from other species, including coral reef organisms, and are similarly useful. The natural fluorescence of such epitope tags allow researchers to optically detect specific types of cells both *in vitro* or even *in vivo* (in the living organism, including mammalians) using fluorescence microscopy.

Hapten A small molecule which can elicit an immune response only when conjugated to a carrier molecule. The term hapten is derived from the Greek *haptein*, meaning "*to fasten.*" Haptens can become tightly fastened to a carrier molecule, such as an antibody, by a covalent bond. In immunoassay techniques, the term hapten depicts a reporter molecule (label) conjugated to an antibody to make it visible. Haptens (i.e., labels such as various fluorophores, enzymes, biotin, digoxigenin, etc.) can be chemically introduced into antibodies via a variety of functional groups on the antibody using appropriate group-specific reagents. The most widely applied principles are covalent haptenylation of amino groups via N-hydroxysuccinimide esters (NHS-ES) or noncovalent labeling using labeled Fab fragments.

Horseradish peroxidase (HRP) an enzyme routinely used in immunohisto-chemistry for labeling antibodies. Histochemichal detection of peroxidase is based on the conversion of aromatic phenols or amines, such as diamino-benzidine (DAB), into water-insoluble pigments in the presence of hydrogen peroxide (H_2O_2).

Immunoaffinity chromatography is a method for isolating an antigen or antibody by exploiting antigen–antibody binding. The antibody or antigen is first immobilized onto a matrix in a column followed by passing a solution that contains the protein of interest through the column. The antigen and antibody may then interact and bind. A buffer that disrupts the antigen–antibody bond is then added to the column, thus releasing the specific protein.

Immunogen any substance capable of generating an immune response (produces immunity) when introduced into the body.

Immunoglobulin See antibody.

Infrared (IR) or far-red light Electromagnetic radiation whose wavelength is longer than that of visible light. The name means "below red" (from the Latin *infra*, "below"). Infrared radiation has wavelengths between about 750 nm and 1 mm. Human vision is insensitive to light beyond ∼650 nm. Therefore it is not possible to view the far-red fluorescent dyes by looking through the eyepiece of a conventional fluorescence microscope.

Micrometer (m) also called a micron It is the metric linear measurement used in microscopy. A micron is 10^{-6}m, and there are 1,000m in a millimeter.

Monoclonal antibodies are antibodies that are identical because they were produced by immune cells that all are clones of a single parent cell.

Monovalent fragments (Fab) of monoclonal antibodies These reagents are prepared from stocks of high affinity antibodies by papain digestion and subsequent purification to yield the approximately 50 kDa monovalent Fab antibody portion of the molecule. Labeled Fab antibody reagents can be coupled to reporter molecules, such as antibodies. This method of noncovalent labeling of primary antibodies is useful in double of multiple immunolabeling when primary antibodies from the same host species are used, such as two or three IgG-type mouse monoclonal antibodies. Masking properties of unlabeled Fab fragments of antibodies directed against imunoglobulins of the species under study are also used for antigen detection on tissues using primary antibody raised in the same species.

Mounting media help the coverslip to adhere to the slide bearing the tissue section or cytological preparation, protect the specimen and the immunohistochemical staining from physical damage and improve the clarity and contrast of the image during microscopy.

Nanometer (nm) A nanometer is 10^{-9}m ($= 10\text{Å}$).

Neofluar objectives are chromatically better corrected than the achromatic objectives (fluorite), but worse than the plan-apochromat objectives. The contrast of the neofluar objectives is however larger than with the plan-apochromat objectives.

Noncovalent labeling of primary antibodies In this method, primary antibodies are noncovalently labeled with a reporter molecule in vitro using as a bridge monovalent Fab fragments that recognize both the Fc and $F(ab')_2$ regions of IgG. The method of noncovalent labeling of primary antibodies is useful in double of multiple immunolabeling when primary antibodies from the same host species are used, such as two or three IgG-type mouse monoclonal antibodies.

Peroxidase See "Horseradish peroxidase."

Paraformaldehyde is a homopolymer of *formaldehyde* with empirical formula $HO(CH_2O)_nH$, where $n \geq 6$. Formaldehyde solutions prepared by dissolving and depolymerization of paraformaldehyde are free of admixtures of methanol and formic acid. Depolymerized paraformaldehyde is useful in enzyme histochemistry, when the preservation of the enzyme activity is of crucial importance, but it has no advantage over formalin solutions routinely used in pathology.

Paratope The part of an antibody that recognizes the epitope, i.e., that part of the molecule of an antibody that binds to an antigen.

Photobleaching (fading) Photochemical reaction of fluorophore, light and oxygen that causes the intensity of the fluorescence emission to decrease with time.

Plan-apochromat and plan-neofluar objectives are corrected for all optical aberrations throughout the visible spectrum from violet to red from center to edges across the entire field of view. They deliver superior image flatness and color reproduction, plus resolving power at the theoretical limit of today's optical technology. These objectives are the perfect choice for multistained, fluorescence specimens and when using all transmitted-light methods. Additionally, high transmission of wavelengths up to the near UV range (360 nm) due to the use of special glass types of low autofluorescence makes plan-neofluar objectives especially appropriate for fluorescence microscopy. Plan-neofluar objectives is the most versatile family of objectives that can be used for all common brightfield and fluorescence microscopy methods.

Primary and secondary antibodies *Primary antibodies* are raised against the antigen under study, whereas *secondary antibodies* are raised against the corresponding IgG species or isotype of the *primary antibody.*

Polyclonal antibodies are multiple antibodies produced by different types of immune cells (different clones) that recognize the same antigen, whereas *monoclonal antibodies* are derived from a single cell line.

Preabsorbed antibody Antibody that has been solid-phase absorbed to remove antibodies cross-reacting with immunoglobulins from other species. Use of such antibodies is recommended when the possible presence of immunoglobulins from other species may lead to interfering cross-reactivities.

Protein A and protein G are bacterial proteins (constituents bacterial cell wall) that bind with high affinity to the Fc portion of various classes and subclasses of *immunoglobulins* from a variety of species. The use of protein A and protein G in immunohistochemistry is based on the same principle as that using secondary anti-IgG antibodies in indirect two-step approach.

Resolution Smallest distance by which two objects can be separated and still be resolved as separate objects. Conventional (light) microscopes are theoretically capable of resolving structures as close as 0.2 m (depending on the wavelength of light).

Spherical aberration Inacccurate focusing of light due to curved surface of lense whereby light rays passing through the lens at different distances from its center are focused to different positions in the Z-axis.

Stokes' shift The difference between the absorption maximum and emission maximum of fluorescent dyes or fluorophores. The Stokes' shift is fundamental to the sensitivity of fluorescence techniques because it allows emission light to be detected against the background coming from excitation light.

Strepavidin A 60 kD extracellular protein of *Streptomyces avidinii* with four high-affinity biotin binding sites. Unlike avidin, streptavidin has a near neutral isoelectric point and is free of carbohydrate side chains.

Tissue microarray (TMA) TMA technology permits to arrange hundreds or thousands of tissue cores (probes), such as clinical biopsies or tumor samples, on a single slide, and then to analyze by a single immuno-staining or in situ hybridization reaction.

Tyramide Labeled *tyramine.*

Tyramide signal amplification This procedure, designated as a catalyzed reporter deposition (CARD) or *tyramide* signal amplification (TSA), takes advantage of horseradish peroxidase (HRP) from an HRP-labeled secondary antibody to catalyze in the presence of hydrogen peroxide the oxidation of the phenol moiety of labeled tyramine. On oxidation by HRP, activated tyramine molecules rapidly bind covalently to electron-rich amino acids of proteins immediately surrounding the site of the immunoreaction. This allows an increase in the detection of an antigenic site up to 100-fold compared with the conventional indirect method with no loss in resolution.

Tyramine (4-hydroxy-phenethylamine, para-tyramine, *p*-tyramine): A pheno-lic amine $C_8H_{11}NO$, a monoamine compound derived from the amino acid tyrosine.

Ultraviolet (UV) light is electromagnetic radiation with a wavelength shorter than that of visible light, but longer than X-rays, in the range 400–10 nm. This part of the spectrum consists of electromagnetic waves with frequencies higher than those that humans identify as the color violet. Therefore it is not possible to view the UV fluorescent dyes by looking through the eyepiece of a conventional fluorescence microscope.

Index